American Miners' Carbide Lamps:

A Collector's Guide to American Carbide Mine Lighting

THE SCRANTON ACETYLENE MINE LAMP

supplies concentrated illumination of high power, at a cost unequalled for economical operation. Our Water Control Valve permits instant regulation of the volume of light. No dirt, grease, soot, sparks or noxious gas. Consumes but one-fifth the oxygen required by oil lamps. Attached to cap or button hole, leaves both hands free. **Write for our Booklet.** It tells all about the world's most practical mining lamp.

Scranton Acetylene Lamp Company
145 Belmont Terrace, Scranton, Pa.

The Chas. Hoppe Co.,
ARTISTIC METAL WORK

Brass, Aluminum and Bronze Castings. Metal Specialties and Platers.

OFFICE, 121 OPERA PLACE. **'PHONE CANAL 329.**

L'ÉCLAIRAGE SPÉCIAL ACÉTYLÈNE

Maisons P. BESSON, André REBATTET, Veuve FERRAND réunies

ANDRÉ REBATTET

Sucesseur

Membre de la Chambre Syndicale de l'Acétylène

66, Rue Claude-Vellefaux— PARIS (X[e])

MARQUE DÉPOSÉE

Téléphone : Nord 27-81 P.B **Téléphone : Nord 27-81**

Spécialité de Lampes de Mines et Chantiers en tôle d'acier emboutie galvanisée. Générateurs intensifs pour Chantiers

CARBURE DE CALCIUM MARQUE TRICOLORE

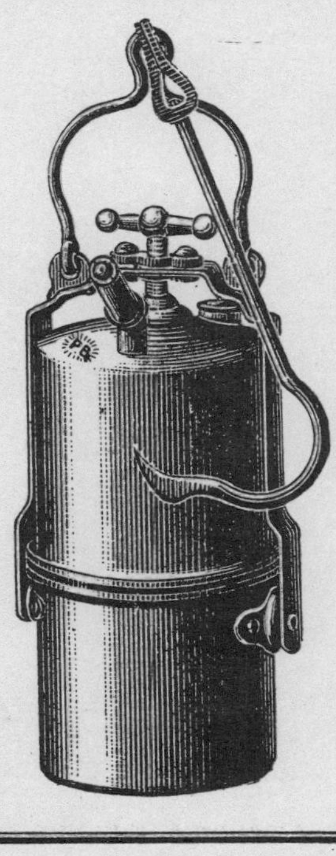

Les plus Répandues

Les plus Appréciées

Fonctionnement simple

Construction Robuste

CATALOGUE spécial franco sur demande

Les Lampes sont en service dans les grandes Mines Françaises et Étrangères, dans les Compagnies de Chemins de Fer, Métropolitain, Travaux Publics, etc.

CATALOGUE spécial franco sur demande

JUSTRITE No. 83
with Bail and Hook

Here is a handy all-'round service lamp—easy to handle —durable—efficient; that's the verdict of the man-on-the-job.

It can be attached to timbers or side of drift. Built substantially to stand up under the hardest kind of slam-bang, day in and day out handling. And like all other Justrites, it gives a clear, steady, brilliant light wherever and whenever it's needed. Write today for prices

Justrite Mfg. Co.
2080 Southport Ave. Chicago, Ill

The Brilliant Search Light

FIFTEEN years on the market, it has stood the test of hunting, fishing trapping, boating, cycling, mining, camping and all kinds of night work. It is the only lamp that is carried on the head, which can be darkened. Has two kinds of lens, one to spread, and one to concentrate the light. One filling of carbide will run the lamp from 4 to 12 hours, according to size of light used. It is so simple that a child can operate it. No danger, absolutely safe. Sold by all dealers. Insist on THE BRILLIANT SEARCH LIGHT—stamped under the generator.

STYLE 1—Single lens, $5.00; Dbl. L., $6.00; Int. L., $6.50.
STYLE 2—Adjustable Candle Flame Burner, add 50 cts.
STYLE 3—Adjustable Flat Flame Burner, add $1.00.

Send for Circular

R. C. KRUSCHKE, K 1810 Superior Street, DULUTH, MINN.

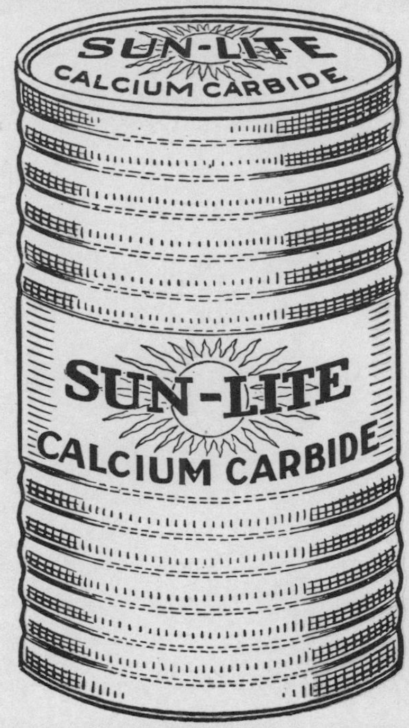

TO KENT SEAVERS . . .

WHO ALWAYS KEPT HIS LAMP BURNING

American Miners' Carbide Lamps:

A Collector's Guide to American Carbide Mine Lighting

By Gregg S. Clemmer

Illustrated by
Wendell E. Wilson

WESTERNLORE PRESS . . . 1997 . . . TUCSON, ARIZONA

SECOND PRINTING
OCTOBER 1997

Library of Congress Catalog Card Number 87-050676
ISBN 0-87026-064-2

PRINTED IN THE UNITED STATES OF AMERICA BY WESTERNLORE PRESS
Post Office Box 35305, Tucson, Arizona 85740

Table of Contents

Introduction and Acknowledgments

"Why would anyone write a book on carbide lamps?" In the course of my research, I have heard that question asked many times. Admittedly a narrow subject, carbide lamps have little appeal outside of a small, specialized circle of collectors, mining memorabilia enthusiasts, and cavers. Yet for those of us who are members of that diminutive yet ardent conclave of acetylene aficionados, any information on these fascinating relics is always welcome.

I saw my first carbide lamp at a Civil War musket match during the Civil War Centennial. Its owner used it to "blacken" the sights of his 1864 Springfield musket. I remember thinking at the time, "What a stinky, sooty device!" Some years later, while flashlight caving in Virginia's Breathing Cave, I encountered several cavers with these lamps mounted on their helmets. Far superior to my cumbersome five-cell flashlight, these carbide lamps provided a broad, even glow of light for the explorers as they passed us in Sand Alley. Thoroughly impressed, I resolved to get a carbide lamp at my first chance. Yet a year would pass before I found one for sale. On a Friday night in the fall of 1968, I bought a new "Justrite" for $3.50 at the Virginia Tech Cave Club meeting. My personal love affair with those "stinky, sooty devices" had begun!

The idea for this book came from a suggestion by a good friend who had found a "Defender" cap lamp and was unable to identify its manufacturer. Realizing that there might be other unusual brands worth collecting, we began a casual collecting rivalry. Always eager to "one-up" each other with a new find, this friendly rivalry continued and then escalated when, between jobs, I began researching lamp patents at the United States Patent and Trademark Office in the hope of gaining an advantage on my acetylene adversary. Slowly, a collection of patents evolved, and upon study, I began to categorize each patent according to the various improvements claimed. Trademarks, while less numerous than patents, provided additional information, particularly concerning manufacturers.

Confronted by my growing collection of lamp data and patents, my friend smugly answered by displaying several recent, unfired finds of his own . . . real lamps! He encouraged me to continue the research and mentioned that one day I should consider getting the material together in book form for other collectors. The idea grew from that conversation and slowly, over the next several months, I began to categorize my information into a working outline.

As a rough plan for such a book took form, it was obvious that any history of carbide lamps should be introduced by a history of the discovery and development of calcium carbide. Thus, the first chapter in the book traces the accidental discovery and subsequent rediscovery of carbide and the early growth of the acetylene industry.

Chapter two is concerned with the first lamps, and follows acetylene to its initial use in underground lighting. Much primary source material was consulted in an effort to document these first, early experiences with acetylene as an underground illuminant. *The Engineering and Mining Journal* and *Acetylene Journal* were of particular value in this regard. Although acetylene lamps burned with a brilliant flame when compared to candles and "sunshine lamps," the first carbide lamps rarely functioned smoothly and hence, lamp reliability in the early days was a continual problem.

Chapter three, Improvements in Lamp Design, explores in detail the various patents and design changes that led to the evolution of a reliable carbide lamp. Highlighted in this section are a number of the most prominent patents. To add a bit of spice and color, I have also included some rather curious and in some cases, bizarre ideas that were patented—all aimed at producing a better lamp.

The use of carbide lamps in the mining industry is discussed in chapter four. Even after the lamp was recognized and adopted by miners for its superior illumination, strong doubts remained about the wisdom of using an open-flame lamp underground. These concerns were to grow and ultimately, by the 1930's, were to lead to the outright banning of carbide lamps in mining. This fact combined with the economic realities of the "Great Depression" spelled doom for the industry. Yet, carbide lamps had managed to find other, albeit smaller markets.

The lamps had been a favorite among sportsmen and hunters for a number of years and by the end of the 1930's, spelunkers or cavers were also adopting them for use in their exploration activities. Thus the chapter Carbide Lamps and Caves, continues the history of acetylene underground. Far from industrial mining, the lamp's service was now primarily recreational, a use that continues to the present day.

The second half of the book deals with the individual manufacturers. For the benefit of collectors, a "thumbnail sketch" is presented on each company. With a gulf of more than half a century to bridge and scant leads to follow, this proved to be a challenging task. Although other collectors around the country provided interesting clues, most of the research was begun by consulting local city and business directories. These sources gave the location of the business, the officers of the company and often a short description of the enterprise. Advertisements, when present, were of notable value. Cross indexing the company officials with their residential addresses provided additional clues. With previous genealogical research experience, I was then able in several cases to contact descendents of these manufacturers. Factory workers, octogenarians all, were found in a few cases and were one of the special treats I enjoyed in the course of the research. Local historical societies and libraries also provided useful information whenever possible. Yet now, at the culmination of this study, I cannot but help sense an incompleteness due to the many remaining, unanswered questions. While not as exciting or popular as political and military history, industrial history can nevertheless contribute an essential chapter to the story of our civilization. I urge others to continue the quest and to shed more light (no pun intended) on these interesting lamps.

It is with this very hope that I included the chapter "Loose Ends" in the book. In the course of researching this book, I encountered many leads and clues that I was unable to resolve. Included in this chapter then are some of the scraps of my research, hopefully presented to stimulate those eager collectors into unraveling some of the more detailed mysteries that remain. Sometimes with only the most tenuous of connections and often playing the "Devil's advocate," I tried to provide enough fuel to "keep the pot boiling" for all those interested in the study and collection of carbide lamps.

In the same vein, I have volunteered a few cleaning and caring tips for collectors. Not intending this section to be an expert guide to lamp restoration and metallurgical science, what I have included is basically common sense. A more precise cleaning guide is included in Dr. Wendell Wilson's *The Nearly Complete Bottom-Guide for Carbide Cap Lamps* cited in the bibliography.

In the last section of this chapter, I bared my soul and revealed my personal tactics and psychology on lamp collecting. To some, this may be more than worth the price of this book. Others may ascribe me as a 20th century carbide Machiavelli! Nevertheless, at the risk of losing future "Gems" (figuratively and literally), I reveal my inner secrets on how to find new lamps.

For the truly dedicated collector, several appendices are included. The first is a fairly comprehensive listing of related patents concerning acetylene mine lamps. Appendix A is further divided into subcategories, based on (1) chronology, (2) technology, and (3) design.

Appendix B is a reluctant addition by myself. Meant only as an indicator, this "rarity scale" is a probable necessary evil in any collector anthology. Everyone wants to know how common or uncommon a particular item is in their collection and lamp collectors are certainly no exception. With numerous collectors pressing for such a listing, it is given with the caveat that it is not meant as a pricing guide (a fool's errand) for antique dealers, and indeed, might change radically if someone were to discover "a hoard" of supposed rarities.

The last appendix is a necessary inclusion as any lamp user knows. (Indeed, some have indicated that Appendix C is the most valuable portion of the book.) Intended to make lamps "user friendly," this "lampitis" section is an accumulated "picking of the brains" of many collectors with years of field experience. It is now included to alleviate the still fickle nature of these otherwise lovable mining relics.

This book exists only through the dedication and contributions of many persons. First and foremost, much credit goes to Dr. Wendell E. Wilson of Tucson, Arizona, for the high-quality art used to illustrate many of the lamps. Locating accurate depictions of many of the lamps proved to be one of the more difficult tasks and I will always be thankful to Wendell for his work and advice.

Lynn Bailey of Westernlore Press believed in the book and gave generous attention, support, and advice throughout the publication process. Many thanks to him and his staff.

A special thank-you to the late George E. Bayles of Cincinnati and his kind daughter Mary, for support, information and gung-ho encouragement in the early stages of the research. Donald Davis of Fairplay, Colorado, caver and lamp collector, generously loaned a large collection of personal papers covering a wide correspondence with many collectors. To fellow Virginia cavers Paula Casale and Doug Molyneaux, I can only say, "Thanks . . . and I wish I had your 'Defender!' "

Deep appreciation and admiration goes to a small but "special" group of helpers. Davis E. Geiger, a former officer of the Ashmead Manufacturing Company, supplied much information from a memory covering 50 years. Mrs. Ben Williamson, daughter-in-law of the Ashmead founder, deserves a special thank-you. Mrs. William B. Jess, daughter of W. Elmer Shanklin, gave interesting clues to her father's and uncles' work, as did Paul Barker respecting his father's early marketing relationship with Frank Guy and the Shanklins.

I certainly owe much to Ted E. Simmons, Sr., son of inventor Samuel E. Simmons. The S. E. Simmons lamp is one of the most scarce lamps and Mr. Simmons generously supplied information to clear up many mysteries. M. K. Phillips, a former lamp assembler, brought a rare perspective to the study as did A. D. Grudgings, a miner from the 1920's.

To a host of cavers and lamp collectors, I can only say that without your help, this book would not be. While I am sure to miss some in this numerous list, many thanks to Ralph Blankenbecler, Larry Peterson, Len Gaska, Mike Puhl, Rick Banning, Paul Kouts, George "Buddy" Stein, Kelly "Deacon" Deem, the late Kent Seavers, Ric Finch, Dr. Dave Thorp, Dave Hubbard, Keith D. Wheeland, Larry E. Matthews, Les Good, Kevin Downey, Ray C. Sease, Andy Yoschak, Don K. Olson, S. A. "Ackie" and Betty Loyd, James Sturgill, Jack D. Ramsdell, Susan Hollar, Emily Davis Mobley, Craig Hindman, John Powers and Henry Pohs.

Helpful assistance was always available at the National Archives, the Bureau of Mines Library—Department of Interior, the United States Patent and Trademark Office, and the Library of Congress.

Historical societies, regional libraries, corporations, and institutions opened many doors and provided numerous referrals. Thanks to Ralph B. Eckley of the *Monmouth Review-Atlas* for information on the "Maple City" lamp. Madelyn Francis of Union Carbide Corporation, New York, provided access to early company files regarding railroad lamps, the scarce Union Carbide lamp, and carbide manufacture. J. L. Ellenberger of the U.S. Bureau of Mines in Pittsburgh offered information on the "Grier" lamp as did James D. Van Trump of the Pittsburgh History and Landmarks Foundation.

Mrs. Ann M. Loyd gave valuable assistance and research leads from her holdings in the Carnegie Library of Pittsburgh. Mrs. Helen M. Wilson of the Historical Society of Western Pennsylvania produced numerous leads and articles on the different manufacturers located in Pennsylvania. William P. Lewis of the Lackawanna Historical Society, Scranton, Pennsylvania, sent information on the "Scranto" lamp. Frank D. Chandler of the Willard Library in Evansville, Indiana, unearthed the remaining scanty information on the "American" carbide lamp. Mrs. Frances Forman of the Cincinnati Historical Society sent information of interest on the "Hoppe Brite-lite" and "Ever-Ready" lamps. Thanks to Mrs. Raymond Horn of the Brick School Museum, What Cheer, Iowa, for information on that town's rare lamp.

I appreciate the efforts of Mara B. Munroe, Oshkosh Public Library for research on the "What-Cheer" and "Oshkosh" lamps. Carol Kelly of the Lincoln Library, Springfield, Illinois, donated many hours in researching Frank Guy, the Shanklins, Meyer Stein, and a myraid of topics concerned with the "Springfield lamp connections."

Thanks also go to Frank D. Paull, Sr. of the Marquette County Historical Society, Michigan, and Eleanor M. Gehres and Augie Mastrogiuseppe of the Denver Public Library. Howard Haynes of the Illuminating Engineering Society of North America, headquarters in New York City, located valuable reference articles on early acetylene testing and lamp development. Mrs. P. A. Black, Premier Lamp and Engineering Company, Limited, Leeds, England, sent current information and some historical references on the "Premier" brand lamp. And kind regards to Cindy Robinson of the Ashland Chamber of Commerce, Kentucky, and Barbara A. Shabroski of the Greater Scranton Chamber of Commerce, Pennsylvania for opening more than a few doors.

National Speleological Society President Paul Stevens provided early support for this project and helped secure access to the Society's records and photos. Other members of the NSS that deserve special thanks are Jack Hess, Jay Arnold, and Mike Dyas. Janet M. Thorne, NSS photo archivist, generously open her files in the midst of moving across the country. For her cooperation I am most grateful. Bradford Cobb, owner of Massanutten Caverns, Virginia, provided history on his cave's illumination by acetylene and is always a treasure to converse with.

Finally, to my wife Heidi, thanks for putting up with stacks of papers scattered all over your house for six years and for believing in my promises to tidy them up . . . soon. Thanks for going along on "lamp hunts" through the hills of West Virginia, Virginia, Pennsylvania, and goodness knows where else. Thanks for enduring the smell of acetylene in your basement when I know it really makes you sick. Thanks for turning a deaf ear to those late night, long-distance phone calls where such arcane terms as "pinch-waist," "burner-tips," "rubber protectors," "polygon-feed," and "brass bottoms" were discussed with such fervor. Now if she would only find me a "Black Diamond". . .

Gregg S. Clemmer

1

The Discovery of Calcium Carbide and Acetylene

Before there could be a carbide lamp, there had to be carbide. The first description of calcium carbide comes from the English chemist, Edmund Davy, in 1836. While developing a method of isolating metallic potassium, he happened upon a most interesting by-product, a dark substance that hardened and crystallized upon drying. Testing its solubility, he was surprised at its vigorous reaction with water and the liberation of a noxious gas. Nevertheless, he failed to recognize the significance of his discovery, gave only token mention to the phenomenon in his reports, and pressed ahead with his potassium experiments. Like so many findings in science, the discovery of calcium carbide and its by-product, acetylene, had been made entirely by accident.

Twenty-six years later, in 1862, Frederic Woehler demonstrated and described the phenomenon that had puzzled Davy. By heating a mixture of charcoal and an alloy of zinc and calcium to a very high temperature, Woehler obtained calcium carbide which decomposed in water to yield acetylene gas. He observed that the gas burned with a brilliant, yet sooty flame. His colleague, Berthelot, in the same year, continued the study and experiments, discovering and documenting other properties of calcium carbide and acetylene.[1]

Still, calcium carbide was procurable only through very tedious laboratory techniques, and with the passage of time, carbide and its gas by-product, acetylene, were all but forgotten. Even as late as the 1880's only a few chemists were familiar with its synthesis and properties.

Then, in 1888, at Spray, North Carolina, Thomas L. Willson began a series of experiments aimed at isolating metallic aluminum. For several years Willson, an electrical engineer, worked toward this goal. In May of 1892, while trying to isolate metallic calcium, he made an unexpected discovery. Using an improved Heroult electric furnace, he passed a current of 2000 amperes with a potential of 36 volts through a finely ground mixture of coke and lime. The resultant product was a dark, melted mass which upon cooling became hard and brittle. It was certainly not the metallic calcium he sought!

According to legend, Willson had the black mass thrown into a nearby stream where to everyone's amazement, the chunk boiled the water with bubbles of some unknown, pungent gas. Someone tried to ignite the bubbles and everyone was startled when a bright, sooty flame jumped from the brook.[2] The experiment was then repeated and this time the dark mass was carefully analyzed and found to be calcium carbide. From the results of this accidental rediscovery, Willson realized he had found a cheap method for producing calcium carbide in quantity.

Willson states that

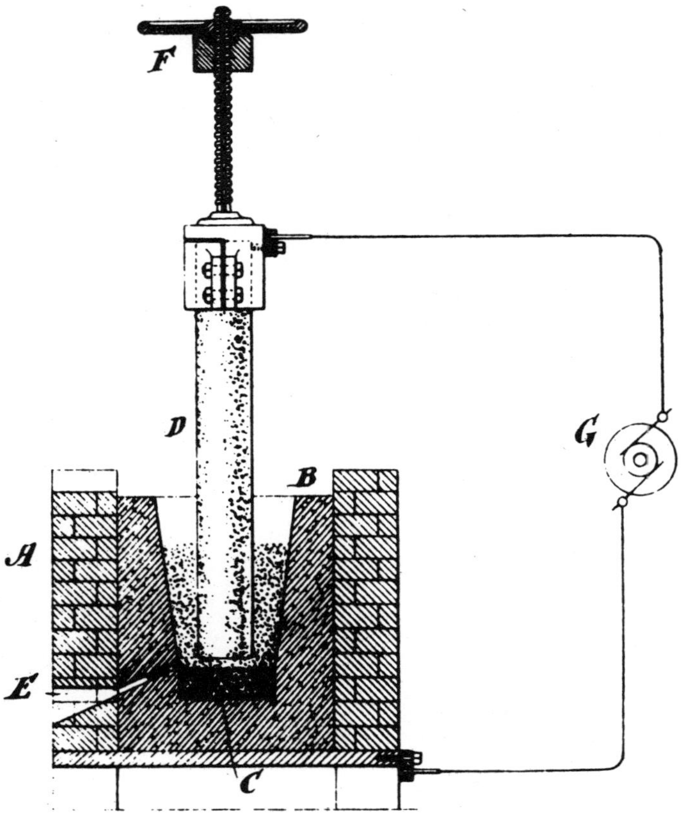

In carrying out my process I proceed as follows: I take very finely divided coke and lime, mechanically reduced to a pulverulent condition, and I mingle them thoroughly mechanically in the proportion of thirty-five per cent of coke and sixty-five per cent of lime. I then subject them to the action of the electric current in a furnace such as that shown in the accompanying drawing, which is a vertical section, and in which—

A, represents an inclosing brick work; B, an internal lining, preferably of carbon, which though preferable, is not in all cases necessary; C, one of the conducting poles, by preference of broken carbon; D, a removable pole of compacted carbon; E; a tap hole for removing the melted product, if desired; F, adjusting mechanism for raising the carbon pole D, and G, an alternating current dynamo which may therefore be made commutatorless.

In starting this furnace, the carbon pole D and the conducting bodies C are connected to the poles of an alternating current dynamo having a mean potential, say of fifty-five volts, and sufficient amperage to produce the amount of material required, having regard to the size of the furnace. Using, for instance, a pole having eight inches on the side, an amperage of about fifteen hundred is desirable. Under these conditions, by reason of the alternation of the current, a feeding in of the mingled lime and carbon between the poles is effected to an extent which does not occur when a direct current is used . . . I have demonstrated that the action is purely a smelting action under the conditions of the process here described. Furthermore, that the block of molten calcium carbide which forms beneath the elevating pole is itself a good conductor of electricity and can be built up from the bottom to any desired height without preventing the operation of the process . . . The liquid calcium carbide thus produced, when allowed to cool, crystallizes . . .

It is essential, in order to produce the new material here described, to have the lime and coke very finely divided. This is done by a powdering machine, and then the powdered lime and coke are thoroughly mingled by a grinder or mixer so as to bring these bodies into most intimate contact. The action of the alternating current is likewise essentially different from that of a direct current, producing a series of explosions, the effect of which seems to be to feed the pulverized material into the arc. This is very important in accomplishing a rapid and uniform conversion.[3]

Subsequent experiments by Moissan and other investigators soon demonstrated the practicality of acetylene as a means of illumination. It was not long before small calcium carbide plants sprang up around the country. Willson's factory at Spray, North Carolina, was just the first of many small carbide manufacturing concerns.[4] By 1908, Union Carbide controlled a large portion of the carbide market with huge production plants at Sault Saint Marie, Michigan, and Niagara Falls, New York.

2

The Early Lamps

After Willson's discovery of an inexpensive method for manufacturing calcium carbide, curious inventors began tinkering and experimenting with a wide variety of carbide gadgetry. Soon, large numbers of acetylene contraptions were being eagerly offered to a public increasingly fascinated by new, labor-saving devices. Prominent newspapers, prestigious magazines, and gaudy tabloids printed glowing reports and advertisements on the new wonder gas, touting everything from the marvelous oxygen-acetylene cutting torch to entire home lighting systems.[1] Indeed, someone even managed to patent an acetylene life preserver and tire pump.[2] Needless to say, in an age of science and machines, carbide-acetylene technology soon found its niche.

It is hard to determine who deserves credit for designing and building the first acetylene light or lamp. Some credit must go, however, to E. Harrogate Gearing for patenting in England on November 16, 1894, a "Lamp for Burning Acetylene." His basic design consisted of "an iron or steel cylinder lamp in which cartridges of carbide coated with geletine, glue, or soluble varnish are decomposed by dropping them into water."[3] This Gearing lamp was at best an awkward device and does not appear to have been intended for underground use.

Others became interested in developing portable acetylene lights. In 1897 American inventor J. C. Gallagher patented one of the first acetylene bicycle lamps, adding improvements with a second patent in January 1898. This first "carbide lamp" of Gallagher's was a simple affair as illustrated in page 16. Already, its similarity in design and technology to the future miner's lamp was unmistakable.[4]

The first known use of acetylene for illumination underground occurred sometime in 1897 when a Mr. H. Hooke, Inspector of Mines in New South Wales, took the opportunity to test acetylene underground while visiting the Millgrove mining field.

> Mr. Hooke has improvised an apparatus including a tank for holding the water, a cage for the carbide, a receiver for the gas generated, a jet to deal it out and a universal jointed reflector to throw the light in any desired direction.[5]

Hooke's curiosity was just the beginning although it is not clear if he continued his experiments. Soon, others followed his lead and carbide began to be earnestly considered for use in the mines.

One of the first serious attempts to use carbide for mine lighting was by the American, Leroy Sunderland Buffington, then living in England. Desiring to incorporate the superior illumination of acetylene gas into an oil safety lamp design, he can be credited with the first attempt at a carbide safety lamp.[6] Interestingly enough, this work emerged before anyone

Fig. 1.

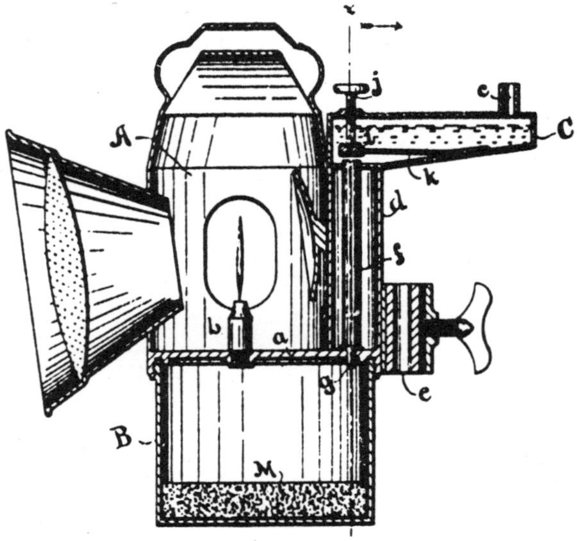

(Above and above right) The J. C. Gallagher Gas Generator for lamps, Patent No. 585,642, June 29, 1897.

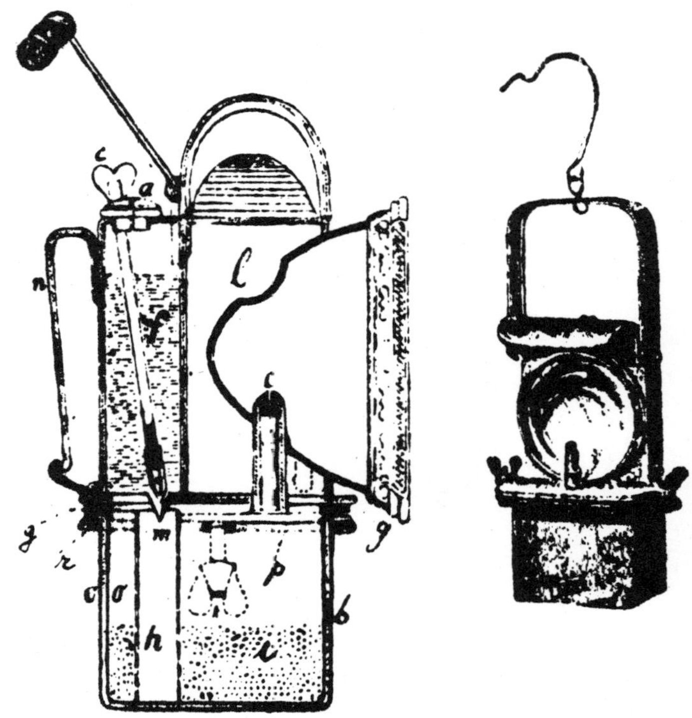

The Velo lamp

Fig. 1.

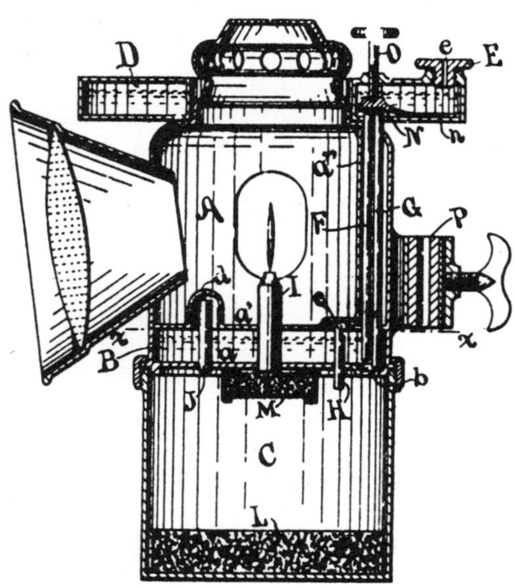

(Above) An Acetylene Gas Generator, patented by J. C. Gallagher, January 25, 1898, No. 597, 900.

hit on the simplicity of the soon popular acetylene cap lamp.

In the November 25, 1899 issue of the *Engineering and Mining Journal,* it was reported that German miners were starting to use acetylene lamps in the New Diepenhock ore mines in the Selbeck district.[7] It is not known who made this lamp, but a report several months later in February 1900, illustrated and diagramed a miner's acetylene lamp that was seeing experimental use in Germany. This lamp was manufactured by the Velo Company of Dresden-Lobtau, Germany, and may have been the first miners' acetylene lamp to be commercially produced. There is no information to indicate how many of these "Velo" lamps were made, or indeed, if any survive today, but in 1899 it was of experimental and therefore, limited use.[8]

The Velo lamp was designed as follows.

> The lower portion of the lamp consists of the carbide receiver, enamelled on the inside and covered by the lid, p, through which projects into the charge of carbide the perforated tube, h serving to convey and distribute the supply of water to the carbide. Between the receiver and the upper part of the lamp is inserted a packing ring of India rubber, which is retained by a steel ring, r, with its two tightening screws, k. At the back of the lamp proper is the water-holder, f, which is filled through the aperture closed by the screw-cap, a. The flow of water to the carbide, through m, is controlled by the regulating screw, c, which can be fixed in any position so as to admit the water drop by drop, or as a thin stream, or shut off altogether. The burner, e, is pro-

vided with two capillary holes, and is interchangeable; and the lamp is completed with a reflector, glass, hood and handles.

Instructions to the user included the following advice.

To keep the lamp in good order the chief essential is cleanliness, The carbide sludge (spent carbide) being soft, can easily be swilled out of the container. Furthermore, it is advisable to only charge the lamp with sufficient carbide to last the time the lamp is expected to be alight, which can easily be regulated, as one oz. of carbide produces gas enough to burn for one hour. A few extra lumps of carbide may be carried in a tin, to serve as reserve supply in case of need.

Interest in acetylene as an illuminant in mines was also being explored in Canada. In an article just four months after the German announcement, *The Acetylene Gas Journal* reported on a silver mine near Ottawa that was generating acetylene outside the mine and piping it through iron pipes and 3/8 inch rubber tubing into the mine. "The consumption takes place through portable burners that can be most conveniently fastened in any crevice of the rock."[9] This was no doubt an interesting and novel use of the gas, but its impractical application in the underground environment must have soon outweighed its limited usefulness.

Meanwhile, in the United States, a new, inexpensive bicycle lamp was being introduced, noteworthy for its simplicity, cleanliness, and efficiency over earlier models. This new lamp, the Baldwin Acetylene Bicycle Lamp, was the first of many to be designed by the prolific, early carbide lamp pioneer, Frederic E. Baldwin.

Living and working in New York City, Baldwin began tinkering with carbide and acetylene and by 1899 had developed a patentable lamp. By May of 1900, his new lamp was being marketed by A. H. Funke, 101-103 Duane Street, New York City.[10] On August 28, 1900, Baldwin obtained his first patent for his ideas and improvements in carbide lamps.[11]

Weighing only 11 ounces and but 6½ inches tall, the Baldwin lamp was designed to burn for four hours, throwing a light for a distance of 50 feet. Built entirely of brass, it was nickel plated and furnished with a German silver reflector which was easily cleaned. Although a bicycle lamp, this design was a significant advance toward the miners' acetylene cap lamp soon to be introduced.

The Baldwin Acetylene Bicycle Lamp was designed as follows.

The carbide, D, rests upon a removable tray provided with a central, perforated tube, C, about which a porous fabric is wrapped. To prevent the carbide, D, from being jolted out of its chamber, a spring-pressed follower is employed, consisting of two, centrally-perforated disks connected by a helical spring. Water is fed to the central tube, C, of the carbide-tray by a small, downwardly-projecting duct provided with a valve, the stem of which is screwed in the top of the water tank, and is operated from the exterior of the lamp. An index-finger, B, is provided to show whether the valve is opened or closed.

The tank having been filled through the opening, A, and the valve opened sufficiently, water trickles into the central tube, C, of the tray, percolates through the porous fabric, and generates gas as it comes into contact with the carbide, D. The gas is filtered through cotton, E, and fed to a burner, located in the focus of a powerful, parabolic reflector. No glass is used; for the pressure of the gas is sufficiently great to prevent the blowing out of the flame.

Every part of the lamp can be easily reached. By releasing the clips which hold the carbide chamber in place, the lamp can be removed without soiling the fingers, merely by lifting the carbide-tray by means of the tube, C. The water-valve can be reached by unscrewing the index, B, and the burner-pipe can be cleaned by unscrewing the reflector.

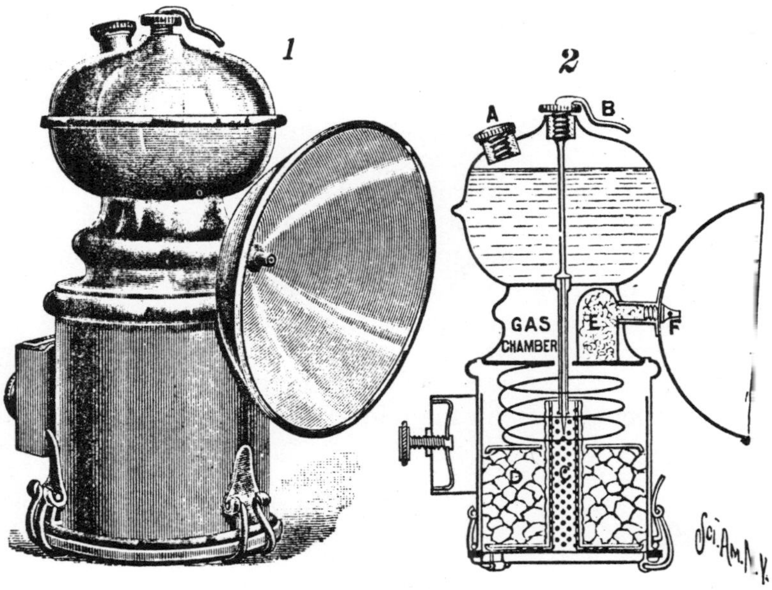

(Above) Two views of the Baldwin Lamp.

It is uncertain when Baldwin began to design and manufacture his lamps for mining illumination, but by 1905, his carbide lamps were being used in both coal and ore mines. A short time later, he became associated with the John Simmons Company as Simmons now began manufacturing the Baldwin and Simmons mine lamps. For the next 20 years, Frederic E. Baldwin continued his experiments with carbide lamps, obtaining more than a dozen patents for his work.

In addition to Baldwin's designs, there were other novel innovations in acetylene lamps emerging. In 1899 Rudolph C. Kruschke, a dealer in guns and hunters' supplies in Duluth, Minnesota, successfully built an acetylene-generator light for hunters and fishermen. His "Brilliant Search Light" consisted of an acetylene generator attached to the waist belt from which a rubber hose piped the gas to a head-mounted reflector-burner. He patented his "Search Light" on October 21, 1902 and as early as 1903, he was recommending it to miners.[12]

Sern P. Watt of the Hine-Watt Manufacturing Company in Chicago, also patented improvements on acetylene bicycle lamps. Predating Baldwin, he did not venture into the mining lamp business until several years after his patents.[13] By 1904 his "Columbia" lamp was being evaluated by mining engineers for use in the mines.[14]

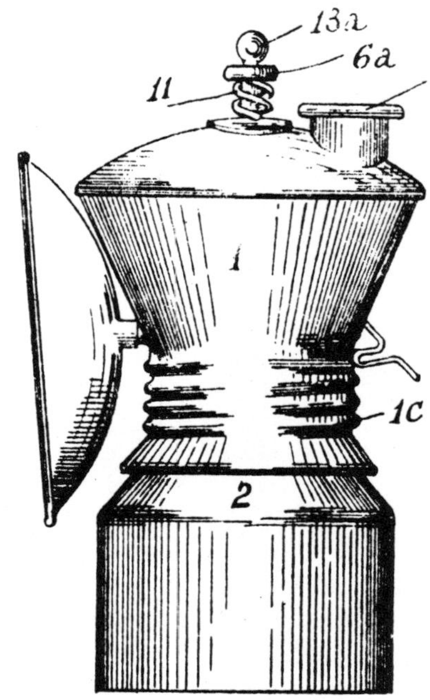

Sketch of the Lamp Bleser used in his improved water-feed design of 1910. Not a manufacturer himself, Bleser probably worked with a "Baldwin" lamp. He may have later worked with Frank Guy and other lamp innovators in Springfield, Illinois.

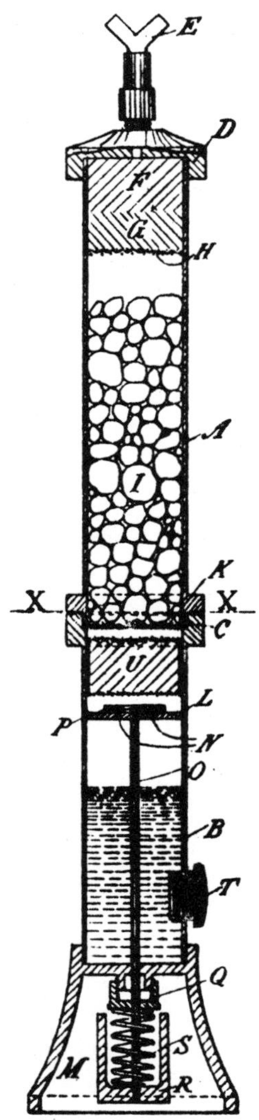

The Marsh "Gas Torch"—The spring-loaded button, R, when pressed with the double burner, E, pointed toward the ground, permitted water inside of B, to pass through the gap created by N, soaking through filter U, and into the carbide, I. Acetylene generated would pass through filters F and G to the burner E for burning. Periodically pressing R kept water going to the carbide and maintained the supply of gas to the burner.

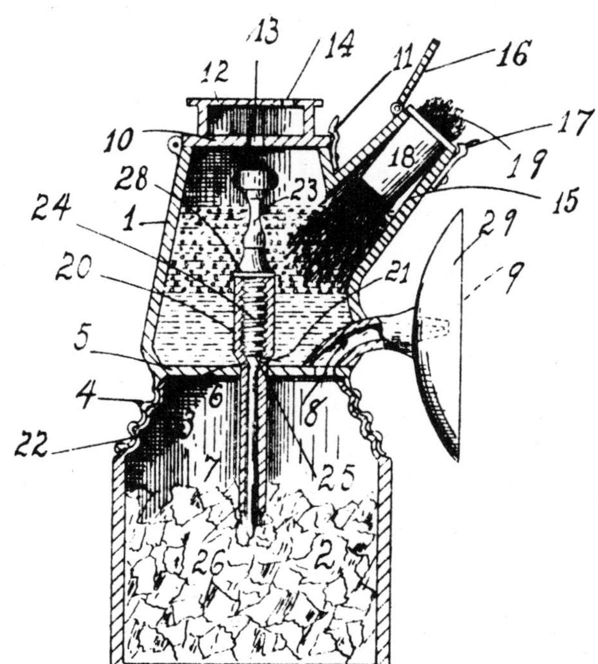

(Above) The Louis Klun oil-acetylene lamp.

In San Antonio, Texas, William H. Marsh patented a device that could best be described as an acetylene flashlight. Cleverly designed, this "gas torch" as he called it, was supposed to be used by railroad personnel to light hidden places when oiling locomotives. It is not known whether Marsh's invention was ever commercially produced, but it can definitely be said that it was the forerunner of the rare Arnold Carbide Candle.[15]

Jacob Bleser, a tool and die maker from Springfield, Illinois, became interested in miners' lamps and patented improvements in the water feed apparatus. His lamp illustration in his Letters Patent dated February 15, 1910, easily shows the advances in design that carbide lamps had made in just a few short years.[16] Some inventors, however, could not totally accept the carbide lamp as the answer to underground illumination. One enterprising individual, Louis Klun of Spaulding, Illinois, designed an oil-acetylene lamp.[17] Using the principle that oil and water will not mix, Klun provided a lamp with two burners: a typical oil wick 19, and an acetylene burner 9. This lamp supposedly had the advantage of permitting the miner to replenish the carbide in his lamp without being in darkness, the oil-wick lamp providing him with adequate light. Again, it is not known whether this lamp was ever produced.*

As acetylene cap and hand lamps proved their superiority over the dim, sooty oil lamps and the still unreliable, expensive electric lights, other carbide lamps from new manufacturers challenged the market. By the start of the World War One, there were more than a dozen manufacturers of hand, cap, or belt-generator carbide lamps. In the years between 1910 and 1934, carbide lamps were to enjoy their greatest popularity in the mines, yielding only then to the inevitable improvements and increased safety of the electric light.

*One of these lamps was found in Illinois in March of 1987.

SOCIÉTÉ INDUSTRIELLE D'ARTICLES D'ÉCLAIRAGE

15, Boulevard Richard-Lenoir, 15 - PARIS

Usines
à **ESQUENNOY**
et à
BRETEUIL-sur-NOYE
(Oise)

LAMPES A ACÉTYLÈNE
PORTATIVES
LANTERNES
LAMPES DE MINES

Marque **"LA PERFECTIONNÉE"**

Fabrique de tous genres d'Appareils, Suspensions, Lustres, Plafonniers, etc. pouvant servir

à l'éclairage à l'Acétylène, gaz, électricité, pétrole, alcool, essence, etc.

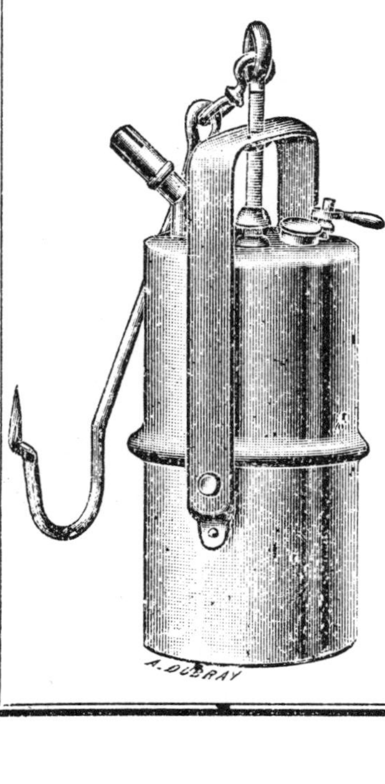

Réchauds et Fourneaux

3

Improvements in Lamp Design

When the first carbide lamps were introduced to the mining industry about 1905, it was apparent that while these lamps offered many new advantages, further improvements would have to be engineered before the lamps could function to their highest potential. To be considered a reliable replacement of the sooty, oil-wick lamps and feeble, tottering candles, the new carbide lamp must ignite easily, burn with an even flame, and provide trouble-free maintenance. It must function safely without danger of exploding, be of rugged construction to withstand the rigors of the underground, and most importantly, be available at an affordable price.

The technology of the miner's carbide lamp at this time was relatively simple. Comparison of lamps circa 1915 with lamps manufactured 50 years later show few basic changes and are a tribute to the inventiveness and sheer hard work of these early designers.

The discussions that follow are a potpourri of lamp improvements, illustrating both important as well as some merely curious developments concerning acetylene lamps. Those interested in exploring further the technology of acetylene lamps are referred to the Patent Index found elsewhere in this book.

A. Flints

One of the first improvements, patented by Alonzo Roach of Linton, Indiana, was the addition of a flint striker on the reflector. This clever gadget made the lamp "self-starting," a time-saving convenience miners, particularly those with tonnage quotas, were quick to approve. Frederic Baldwin patented additional improvements to Roach's idea on March 7, 1916, claiming "an acetylene miner's lamp having an igniting device which can be easily removed for adjustment and repair and can also be securely and firmly locked in proper position when required for use."

B. Shock wave, blast, or breeze protectors

One of the miners' first complaints about carbide lamps was how easy it was for a lamp to be extinguished by mine breezes and shock waves from blasting. Frederic Baldwin was the first to patent any remedy to this problem. His Letters Patent No. 1,083,427 dated January 6, 1914, utilized a donut-shaped rubber bag situated securely behind the reflector. Inflated by acetylene generated inside the lamp, it was therefore sensitive to any change in outside pressure.

Several months later, Frank Guy, a miner from Springfield, Illinois, patented a pressure device consisting of a rubber bulb conspicuously mounted on the side of a lamp. Guy, in the same patent, claimed improvements in the water-feed valve. This feature overshadowed his pressure device and was the princi-

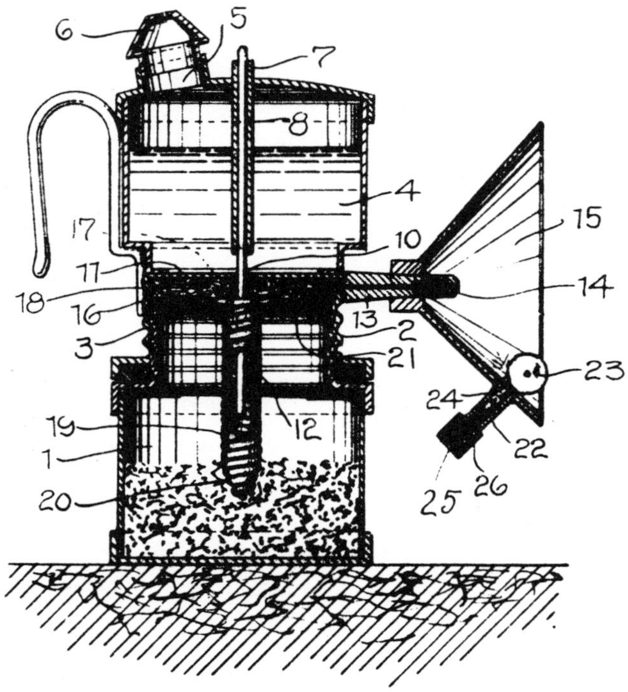

(Left) Cross-section of typical carbide lamp, circa 1915. This drawing is taken from Alonzo Roach's patent No. 1,068,931 dated July 29, 1913, in which he claims the first reflector-mounted rotary flint striker. The sketch diagrams the basic parts of a miner's carbide lamp showing little functional change since 1915. Water located in the top chamber, 4, drips through the water feed valve, 7, 8, 10, 12, onto the calcium carbide in the lower container, 1. The resultant gas generated, acetylene, is pressured through a filter, 18, usually of felt, held in place by a filter retainer, 21. The acetylene is then forced through a small pipe, 13, to a burner tip, 14, mounted centrally around a shiny reflector, 15. A flint striker assembly, 22-26, mounted in the reflector is then available for manual ignition of the flammable acetylene. A hook or similar attachment device, 17, is available for cap mounting, holding, etc.

pal technology behind the Guy's Dropper of the Shanklin Manufacturing Company.

Later, Baldwin received two other patents on his inventions in this area, the most sophisticated being No. 1,237,450 dated August 21, 1917. Basically a variation of Guy's design, it featured a larger bulb of more durable construction connected to the gas gathering area directly behind the burner tip. This bulb, 19, protected the pliable rubber membrane, 13.

In addition to sudden pressure waves from blasting, mine breezes created problems for the new lamps. A. B. Mitchell patented an idea whereby the flame was shielded by convection currents created by the heat from the flame and shunted around and through a series of holes in the reflector.

John M. Brock of the Dewar Manufacturing Company in Brooklyn, New York, probably designed the first flame protector, filing his patent application in October, 1915. Patented June 11, 1918, his idea was to screw an extension to the burner tip. He refined his invention with patent No. 1,461,572 dated July 10, 1923. This tip protector was commonly seen on most Dewar lamps.

Harry F. Roberts from Kay Moor, West Virginia, probably developed the simplest flame or tip protector. His idea was to screw an angled strip of metal, holed correspondingly to the burner tip, to the reflector. Like Brock, his patent hinged on the fact that as

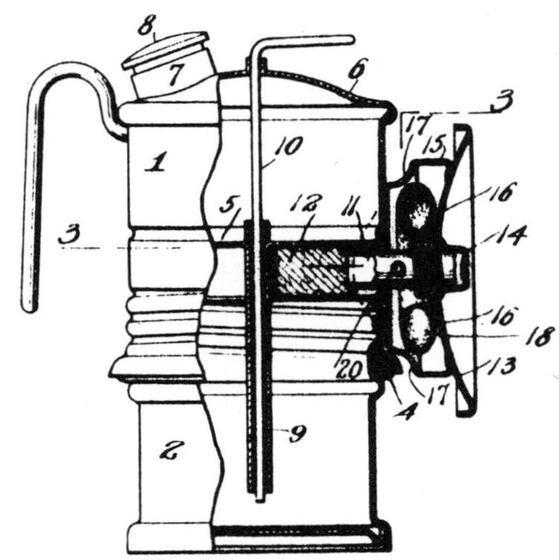

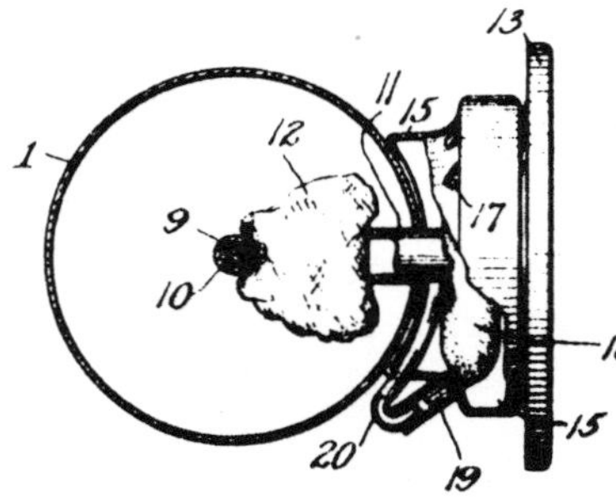

(Left) Diagrams from Baldwin Patent No. 1,083,427. "If, now, a sudden increase in the atmospheric pressure outside the lamp should take place, any tendency of this pressure to drive back the outflowing gas in the burner tube, 11, will be overcome by the compression of the elastic bag, 18, which will result in an equally sudden increase of pressure within the burner tube just back of the burner, thus balancing the pressure inside said burner tube with the pressure of the air outside the lamp."

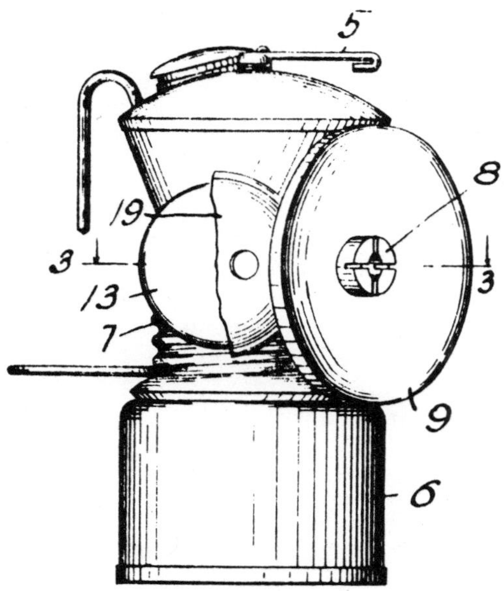

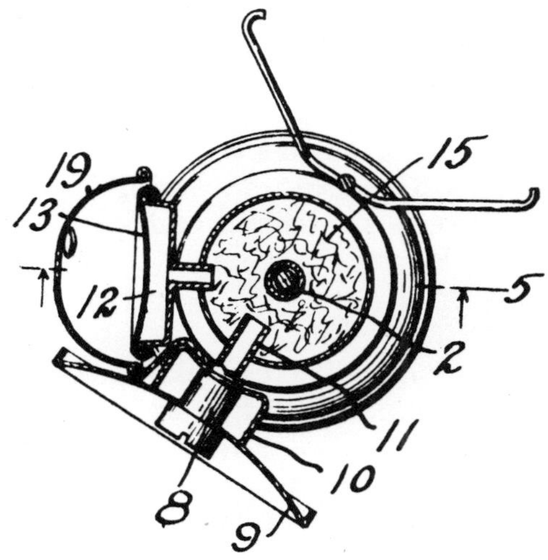

(Left and above) Frederic Baldwin's "Barometrically Protected" Lamp circa 1917 with accompanying horizontal cross-section. It is not known if many of these lamps were ever produced and any in existence would certainly be rare.

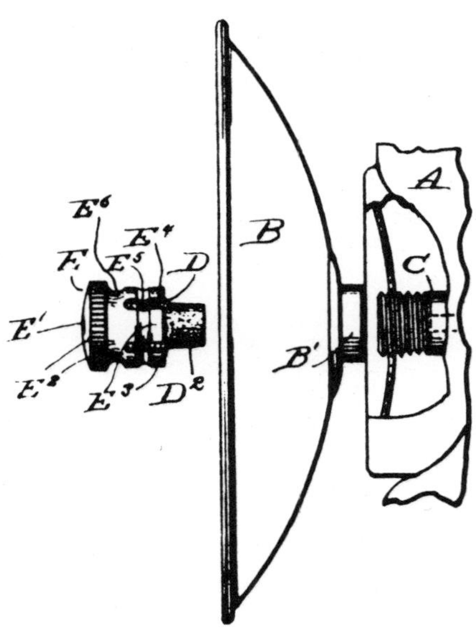

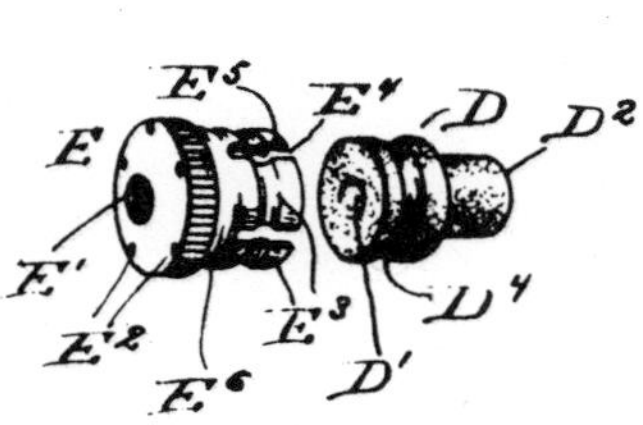

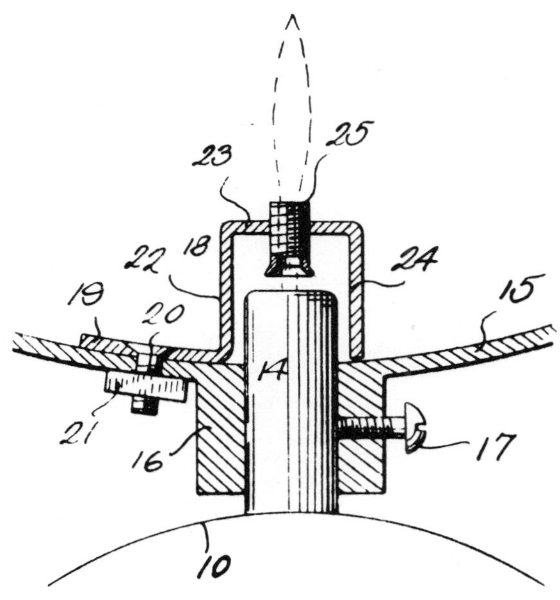

(Above) The Brock-Dewar Tip Protector. The "lamp A is provided as usual with the reflector B, the burner tube C, the burner tip D, and the protective cap E, the latter covering the burner tip and constructed to form a flame-surrounding chamber into which the burner orifice opens."

(Above) Close-up of burner tube, reflector, and flame guard. Flame extends from burner tube, 14, through the large diameter tube, 25. Protector is mounted to reflector, 15, by screw, 20, and nut, 21, through holed strip, 19. Reflector is mounted to burner tube by set screw, 17. 10 is the body of the carbide lamp. Patent No. 1,280,849 dated October 8, 1918.

long as the origin of the flame at the burner tip was protected from breezes, the flame could not be easily blown out.

Flame protectors such as the Roberts design gave way easily to the better working varieties distributed later by the major manufacturers. Two good examples of these accessories were the Justice No. 170 and No. 180 flame protectors, readily available for a number of years.

C. Water valve and water-feed apparatus

One of the biggest technological problems confronting the carbide lamp designers was that of establishing a satisfactory water-feed mechanism. This was essential in order to insure a steady, smooth flow of acetylene to the burner tip. As early as 1900, Baldwin recognized the problem.

> The water valves . . . have not met satisfactorily the requirements of practical use. It is generally true of all of them that the tendency to become clogged up by the sediment and impurities in the water soon renders them inoperative. If a needle-valve is employed, the acid in the gas is apt to corrode the contacting surfaces, so that they leak, while if some ordinary and simple form of cock is used it is very difficult to regulate the flow of water through it properly.

These problems were not easily surmounted and a number of ideas were tried before adequate results were realized. In addition to the popular needle-valve device, there was the patented design of Samuel E. Simmons of Litchfield, Illinois. James Taylor of nearby Peoria utilized a "V" shaped water chamber that extended into the carbide container. Others, such as machinist Jacob Bleser, designed lamps that were simply variations of the needle-valve. Bleser cited his claims by stating that

> The purposes of this invention are to provide a valve body of improved construction, to provide a valve stem easily removable for the purpose of cleaning the valve, to provide means for connecting the feed-tube with the valve-body, to provide means for cleaning the feed-tube without detaching the valve or the valve stem. . . .

Even with all of these various "dropper" mechanisms, no easy remedy was devised. If a particular design solved a problem it only seemed to open the door to another difficulty. The needle-valve was still the best that technology had developed, but even this was lacking. Baldwin, just prior to the introduction of his "ZAR" brand carbide lamp in 1914, wrote:

> When manufacturing these lamps in large quantities however, it has been found impossible to proportion the two diameters (of the valve space and the needle itself) so that the flow of water should be even approximately correct. This difficulty arises from the necessity of using a fairly long water tube in order to maintain the desired water column pressure and any slight deviation in the diameters of the wire and tube cause a very marked variation in the flow of water, owing to the amount of friction on the water induced by the length of the restricted water tube.

Baldwin attempted to solve these problems with further modifications, patented No. 1,101,982 issued

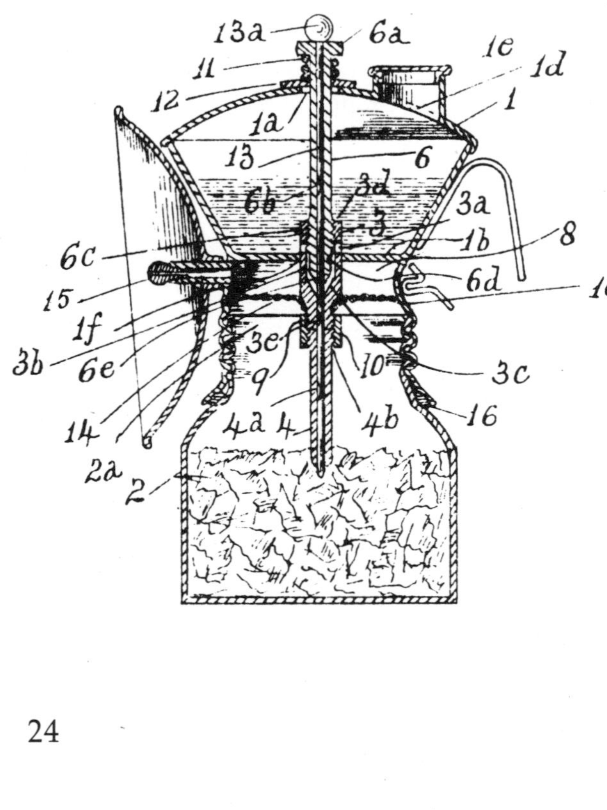

(Left) Diagram of Patent No. 949,349 dated February 15, 1910, to Jacob Bleser. The valve body 3 has internal threads 3d, openings 3a, a central chamber 3b, a tapered valve seat 3c, at the bottom of the chamber 3b, a downwardly extending reduced screwthreaded member 9 and a central vertical hole 3e in which the needle 13 fits loosely. The feed tube 4 has a central bore 4a in which the needle 13 fits loosely and has at its upper end a screw-threaded part 4b. A coupling 10 screws onto the member 9 of the valve body and the screw 4b of the tube 4 and connects the tube with the valve body. The valve stem 6 has a central bore 6b, a screw-threaded part 6c screwing into the screw 3d of the valve body, a reduced part 6d within the chamber 3b of the valve body and having its lower end 6e tapered to fit on the valve seat 3c; and has at its upper end a knurled handle 6a by which the valve stem may be turned to screw it into the valve body, or to unscrew it from the valve body. The washer 12 fits around the valve stem and covers the opening 1a. A coiled spring 11 surrounds the valve stem between the handle 6a and the washer 12 and acts to hold the washer firmly over the opening 1a, so as to prevent the escape of water through the opening. The needle 13 fits loosely within the bores of the valve stem, the valve body and the feed tube and has at its upper end a knob 13a for convenient handling of the handle.

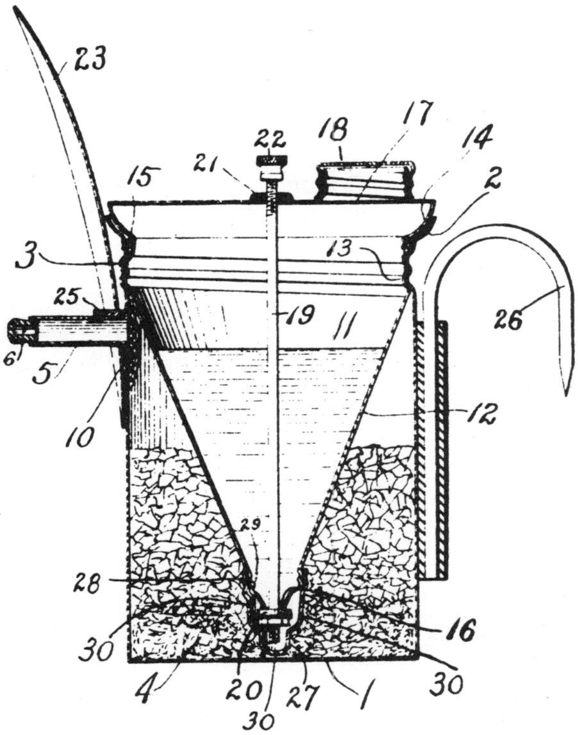

(Above) Diagram of Patent No. 996,950 dated July 4, 1911, to James Taylor. Note the "V" shaped water chamber. Supposedly, knob 22 acting through control rod 19, opens or shuts water valves, 30, allowing water to seep into the carbide 4. The efficiency of this design, however, is questionable.

June 30, 1914 and No. 1,237,449 issued August 21, 1917. Augie L. Hansen, an inventor for Justrite Manufacturing Company in Chicago, also worked with the needle-valve, patenting a knotched control knob, patent No. 1,084,145, in January, 1914. Nevertheless, these and other fine-tune adjustments in the needle-valve still did not lead to a design rugged enough for the miners' needs underground.

Finally, it indeed took a miner to design a strong, dependable water-dropper mechanism. This miner, Frank Guy of Springfield, Illinois, filed his application on March 1, 1911 "to provide a novel water-supply device that is so constructed and valved as to insure a reliable feed of the water into the carbide, while at the same time being protected against clogging." Patented on May 26, 1914 and improved in subsequent patents, this water control mechanism was incorporated into a new miner's carbide lamp bearing its inventor's name, the "Guy's Dropper."

The addition of the "valve head or button" in Frank Guy's patent now gave more control to the water-feed valve. Both the speed and quantity of the water dropping to the carbide could be regulated evenly. On September 19, 1916, Guy was issued patent No. 1,198,537 for further improvements in his dropper design.

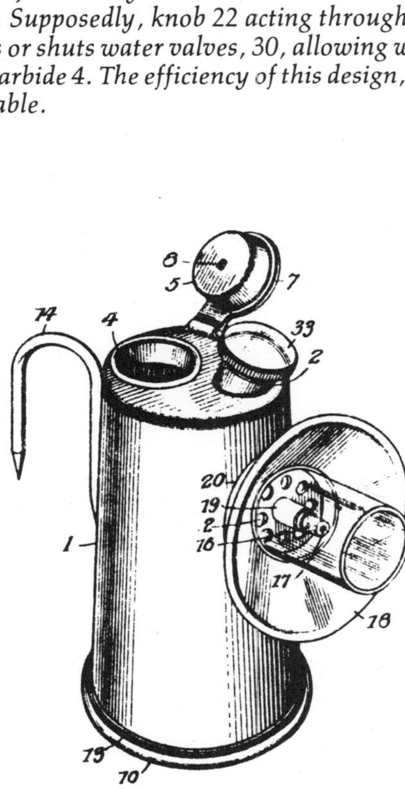

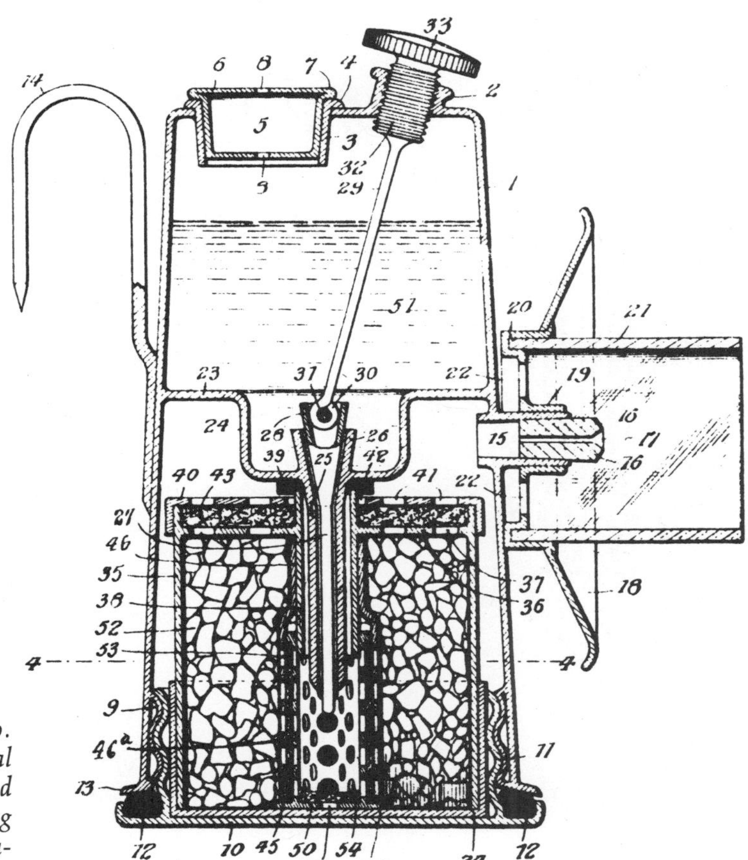

(Above and at right) Samuel E. Simmons' Patent No. 976,611 dated November 22, 1910, including an external view of the lamp. Through threaded control knob 33, rod 29 lifts or lowers water plug 27 thus opening of closing access to tube 25. If opened, water passes through perforations 46a to the carbide 52.

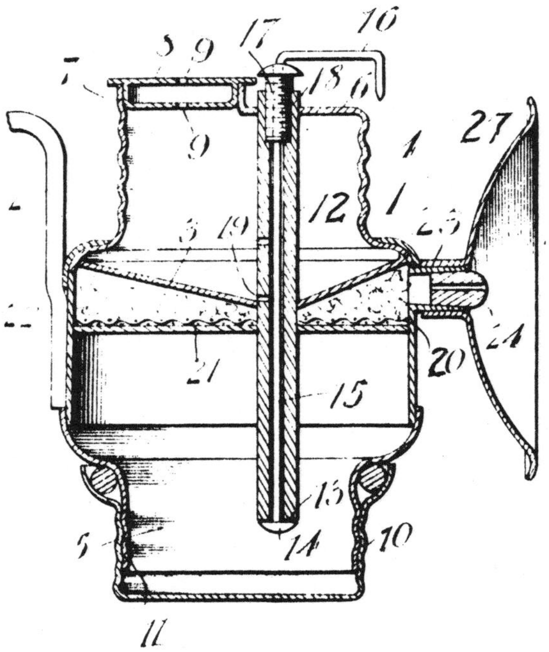

(Left) Diagram of Patent No. 1,097,716 issued to Frank Guy, May 26, 1914. "Water is supplied to the chemical in the chamber 5 by a water-supply device consisting of a tube 12 arranged longitudinally of the casing, and having its upper end projecting through the casing top 6, while its lower end lies within the gas chamber 5, and is provided with the bottom outlet opening 13, which is covered and uncovered, more or less, by a valve head or button 14, carried at the lower extremity of an adjustable valve stem 15 extending longitudinally through the tube 12 and provided at its upper end with an exterior operating handle 16 . . . the valve head or button has a revolving or grinding action at the lower end of the valve tube which insures the maintenance of the water supply for the carbide chamber." Note the screw-on carbide chamber and general odd shaped-design of the lamp.

With this patent, the water feed problem in carbide lamps was essentially solved. This valve was used extensively by Shanklin Manufacturing Company in the production of "Guy's Droppers" and continued later by the Universal Lamp Company. Universal's popular "Auto-Lite" did not use this valve, but instead employed a simple ball plug or valve with an extended mixing rod, drawing from Guy's 1914 patent and taking the rod from Augie Hansen's patent of October 24, 1916.

Subsequently, there were patents by other manufacturers, slightly improving the "Dropper" with various gadgetry. Meyer Stein, manufacturer of the "Springfield" brand lamp, patented a removable and therefore cleanable control needle. Refining Stein's patent, Ben Williamson, manufacturer of the "Buddy" and "Elkhorn" brands was issued patent No. 1,716,809 on June 11, 1929.

Having solved the problem of maintaining an even flow of water to the carbide, most inventors worked on another related problem.

> It is well known . . . that efflorescence [*sic*] results from the contact of the water with the carbide, causing a moist ash to accumulate in and around the lower end of the tube, or around the valve at the lower end of the tube as the case may be, and when the ash dries it forms incrustation which often completely chokes the tube so that water cannot pass through it; or surrounds the valve and causes it to stick on the valve seat thus stopping the flow of water through the tube so that it is necessary to frequently rotate the valve to break up the incrustation and restore the feed.

Augie Hansen of Justrite, designed several mechanisms in an effort to solve this problem at the same time patenting an interesting spiral valve stem for the water-feed. This spiral valve stem had been preceeded by an even more sophisticated spiral water feed system, intricately designed and beautifully unique. Unfortunately, it was too delicately complicated for adequate use in the mines.

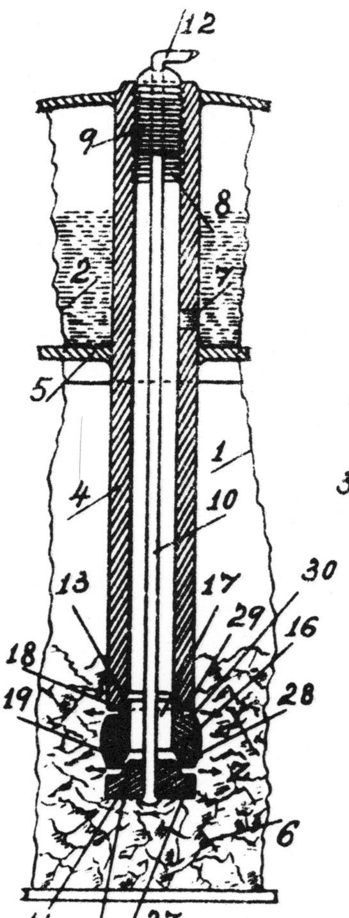

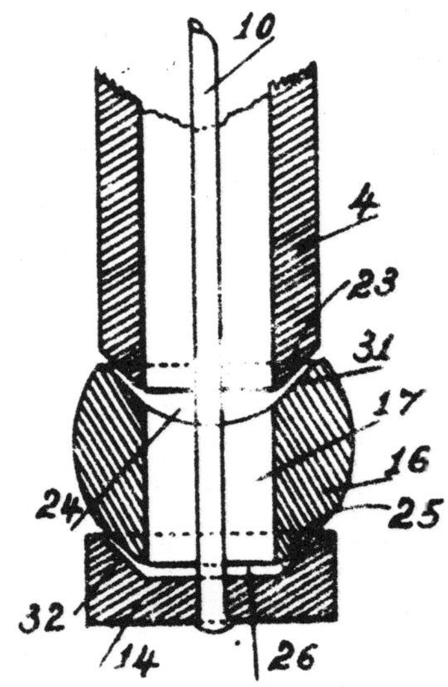

Fig.4

(Above) Cross-sectional diagrams of Frank Guy's Dropper Patent of 1916, No. 1,198,537. In the left figure, water 8, entering hole 7, fills interior of valve flowing to carbide 6, when handle 12, is turned lowering plug 14, and creating spaces 18, 19, above and below hexagonal plug 16.

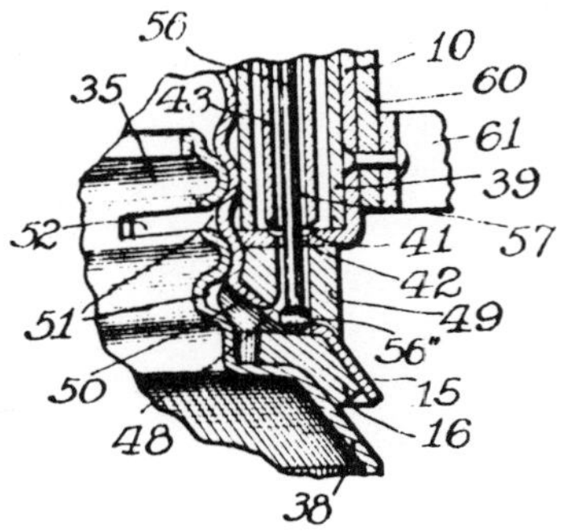

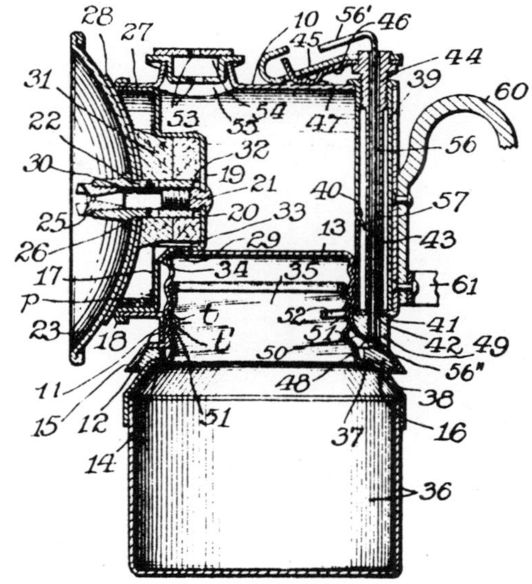

Hansen, a prolific inventor and pioneer of carbide lamps,later patented improvements in this spiral water feed (No. 1,310,529 dated July 22, 1919) but increasingly spent more of his energies on other technological problems concerning the lamps. Interested with the encrustation-blockage problem, he received three patents for a water sprinkling, carbide mixing device engineered to eliminate the difficulty. Progressing from his 1915 patented design, note the spiral valve stem employed in the water feed, illustrated below.

A much more practical solution to the "encrustation" problem was the invention of Patrick McDonald of Westville, Illinois.

Other attempts to prevent encrustation involved a coiled carbide mixer capable of spreading water evenly. First patented by Henry Kerst on February 6, 1912, this coiled spreader was improved by John M. Brock and used widely by the Dewar Manufacturing Company in their lamps.

Of the water feed devices discussed so far, all have depended on gravity, i.e. the passive flow of water to the carbide by mounting the water chamber superior to the carbide-generating chamber. Yet, there were several attempts to use gravity plus a head of positive pressure to insure a steady, controlled flow of water to the carbide.

Another oddity was Meyer Stein's double dropper system. This was designed to accommodate two separate carbide generating chambers, one to act as a reserve when the other was being refilled. In this way, no lamp would ever have to be extinguished during a refilling.

In addition to these varied attempts to improve the needle-valve, water-feed mechanisms in carbide lamps, either by modification or alternate designs, there were many other similar ideas introduced and patented. For one reason or another, most of these patents were simply stepping stones, direct or indirect, to a simpler, yet more efficient solution. From primitive needle valves to intricate, spiral dropper systems to decrusting devices to positive pressure pumps and shunts, the technology of the miner's

(Above and above left) Diagrams of Hansen's intricate spiral water feed system, Patent No. 1,162,915 dated December 7, 1915. ". . . a water duct 39 extends vertically through the water container shell 10 adjacent the rear wall thereof, this tube having the spiral slot 40 for the passage thereto of water from the water container. In the base of the shell 10 and concentric with the tube 39 is the valve seal 41 surrounding the outlet 42. The lower end of a tubular valve stem 43 is tapered to cooperate with the valve seat to control the flow of water through outlet 42, the head 44 of the valve stem having threaded engagement with the upper end of the duct 39 in order that when the stem is turned the tapered end thereof will be adjusted with reference to the valve seat. An actuating lever 45 extends from the head 44 and is locked in adjusted position by the cooperation of a spring arm 46 with protuberances 47 on the shell 10. Directly below the valve outlet 42 the shoulder 14 of the cap 13 has the outlet opening 48, and a short tube 49 interposed between this shoulder and the base of shell 10 forms a connecting passageway between the valve outlet 42 and outlet 48. In order to prevent closure of the outlet 48 by the washer 16 when the carbid container is screwed into the cap 13, the wall section 50 adjacent the opening is deflected upwardly a distance . . . the arrangement involves the provision of a groove or channel between the neck 35 of the carbid container and the cap 13, one end of the groove or channel communicating with the water outlet 48, and the other end communicating with the carbid container. The groove or channel may be spiral and formed by so designing the threads that the ridge of one will not extend entirely to the base of the other . . . the lower end of the passageway 51 is adjacent the water outlet opening 48 and the upper end of the passageway communicates through slot 52 with the interior of the carbid container.

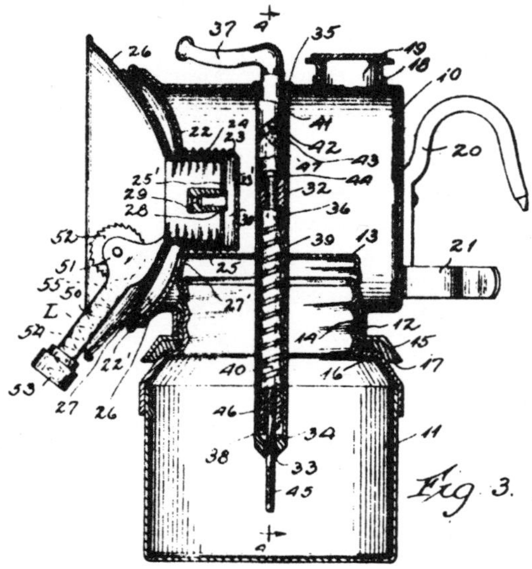

(Above) Diagram of Patent No. 1,202,514 dated October 24, 1916. The Hansen spiral water valve stem. Note the interesting reflector attachment to the lamp body. Justrite Manufacturing Company, Chicago, Illinois.

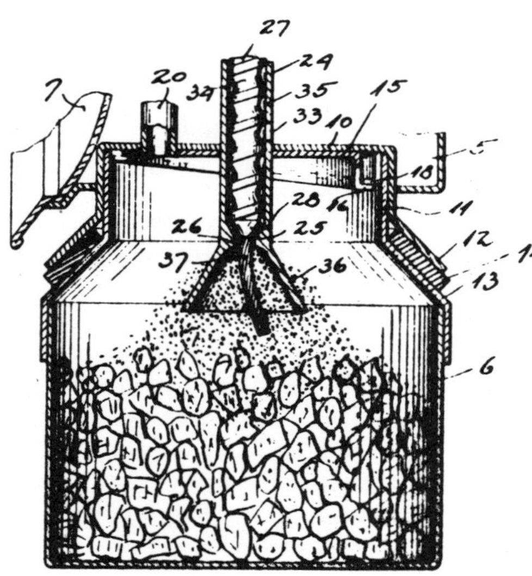

(Above) Diagram of Patent No. 1,224,537 dated May 1, 1917. The Hansen spiral water valve stem with angled water spreader and carbide mixer. Justrite Manufacturing Company, Chicago, Illinois. In Patent No. 1,233,663 dated July 17, 1917, Hansen straightens the carbide mixer, providing deeper agitation of the carbide.

carbide lamp water-feed system challenged the best engineers in the industry. With Guy's patents in 1914 and 1916, the dropper system was essentially complete. With William J. Frisbie's patent for the Justrite Company in 1922 the dropper was successfully combined with a workable encrusting device.

Within a short time, most of the Justrites were incorporating this design or a similar variation, and today, the majority of lamps in existence possesses a variation of this dependable, rugged water-feed mechanism.

D. Acetylene filters, felts, felt holders

It was readily apparent to the early developers of carbide lamps that some form of filtering system would be necessary before freshly generated acetylene could be burned cleanly at the burner tip. Frederic Baldwin and other early lamp designers struggled with a variety of filter materials including wire mesh, felt, cotton and burlap, with varying degrees of success. Peter Togleson of Deepwater, Missouri, stated the object of these devices in his patent dated September 13, 1910.

> . . . to produce . . . a novel arrangement of sifters or screens interposed between the carbide holder and the burner for consuming the gas where-by the clogging of the conduit or pipe and the burner tip is effectively prevented, means being also provided for rendering the parts just described readily accessible for the purpose of removing deposits of foreign matter. . . .

Togleson's patent was simply a description of a double layer of felt interspersed by a gas collecting

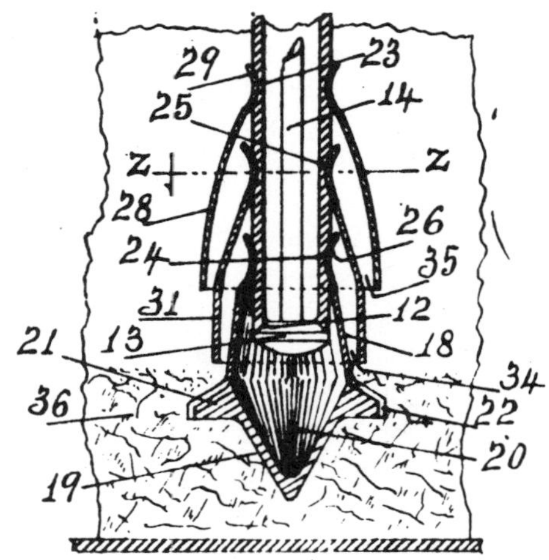

(Above) Diagram from Patent No. 1,356,711 dated October 26, 1920, to Frank Guy. Designed as a carbide agitator and sophisticated water spreader, the complexity of the mechanism doomed its usefulness. Water, if blocked by crusts, etc., at 20, could rise and flow into space 34, or if this is blocked, to space 35. Shanklin Manufacturing Company, Springfield, Ill.

space, possibly an improvement, but not a significant advance in filter design. If the water-feed apparatus was a chief cause of lamp trouble, a properly functioning gas filter could prove just as fickle. By 1915, the state of the art could best be termed "tolerable."

> . . . gas generated . . . is filtered by being passed up either through a single thickness of wire gauze to the burner or through such wire gauze and a felt or porous filter disposed between the gauze screen and the gas outlet. In lamps using a wire gauze screen of a single thickness it has been found that the gas is so imperfectly filtered that the burner becomes quickly clogged. The extreme delicacy of an abnormally fine mesh wire screen, its expensiveness, and the fact that it is subjected at times to very considerable gas pressure, makes it impractical to use screens finer than sixty mesh and such a screen will not by itself properly filter the gas and free it of dust and ash. Where filtering material, such as felt, cotton fiber, or the like is interposed above the wire screen, it has been found that such material tends to quickly become choked with the dust so that it will not deliver a uniform and regular supply of gas to the burner . . . such porous filters require constant attention.

Other inventors advanced more sophisticated filters. Take, for instance, the unique "plug" filter patented by A. H. Mollman, Jr.

Joseph Messina of Benld, Illinois, patented a spiral wick or felt filter in 1917. Joseph Menapace of Latuda, Utah, embellished Messina's design with ideas of his own and was issued patent No. 1,552,218 on September 1, 1925. Although probably never manufactured in any quantity, these interesting designs reflect the intricate attention given to the filter problem.

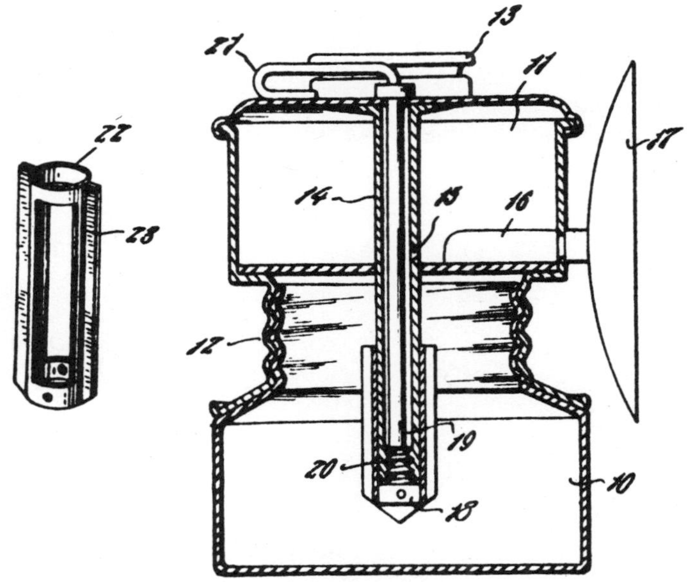

(Above) Diagram of Patent No. 1,712,392 dated May 7, 1929. McDonald's simple carbide mixer as attached to the water feed control tube.

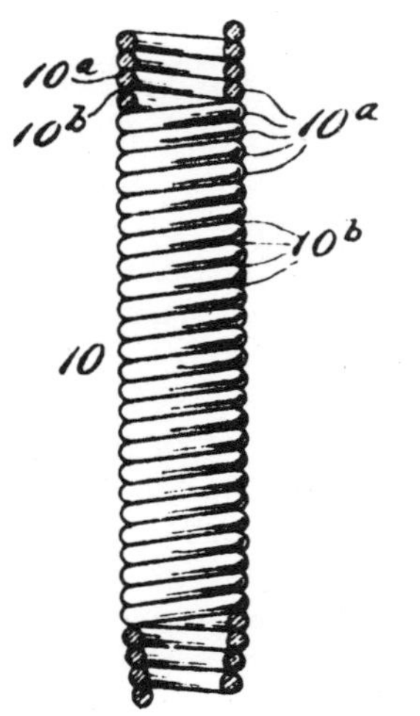

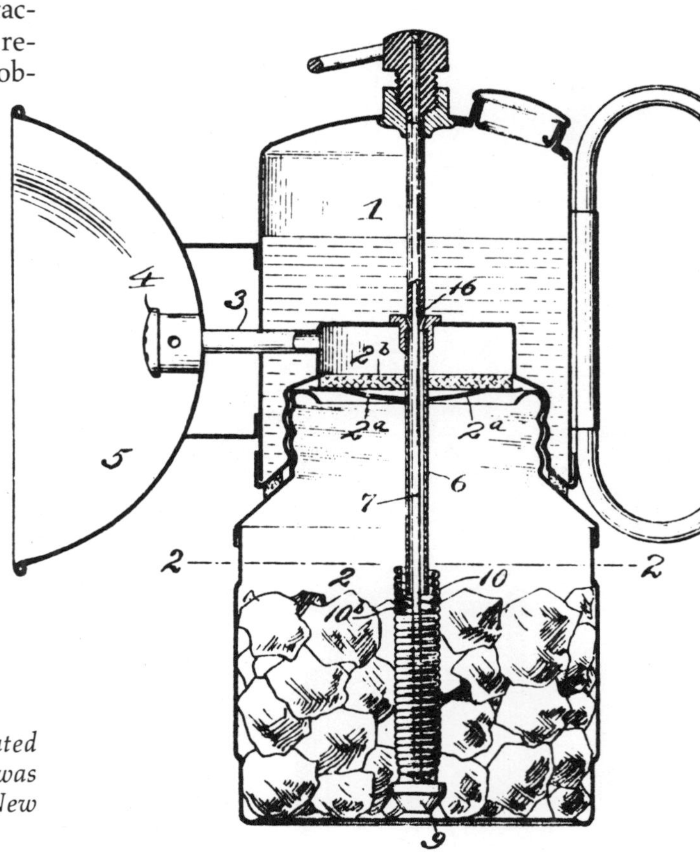

(At right) Diagram of Patent No. 1,605,505 dated November 2, 1926. The coiled water spreader (above) was widely used by the Dewar Manufacturing Company, New York, New York.

(Below) Diagram of Sherman Horsley's Patent No. 1,312,731 dated August 12, 1919. A piston pump is mounted to the lamp to pressurize the water chamber and force water onto the carbide. In addition, Horsley added a gas shunt 29 for excess acetylene to be piped back to the water chamber to maintain a constant head of pressure.

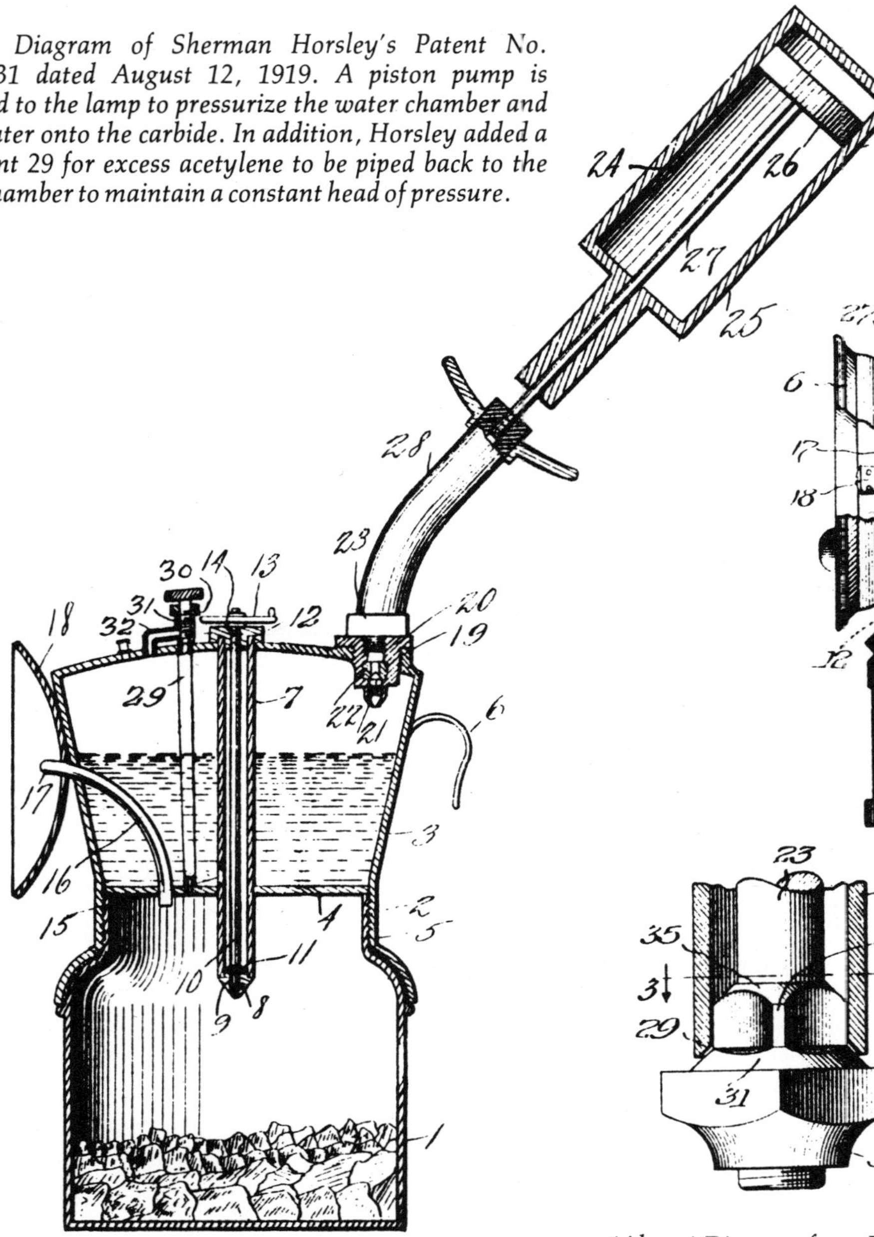

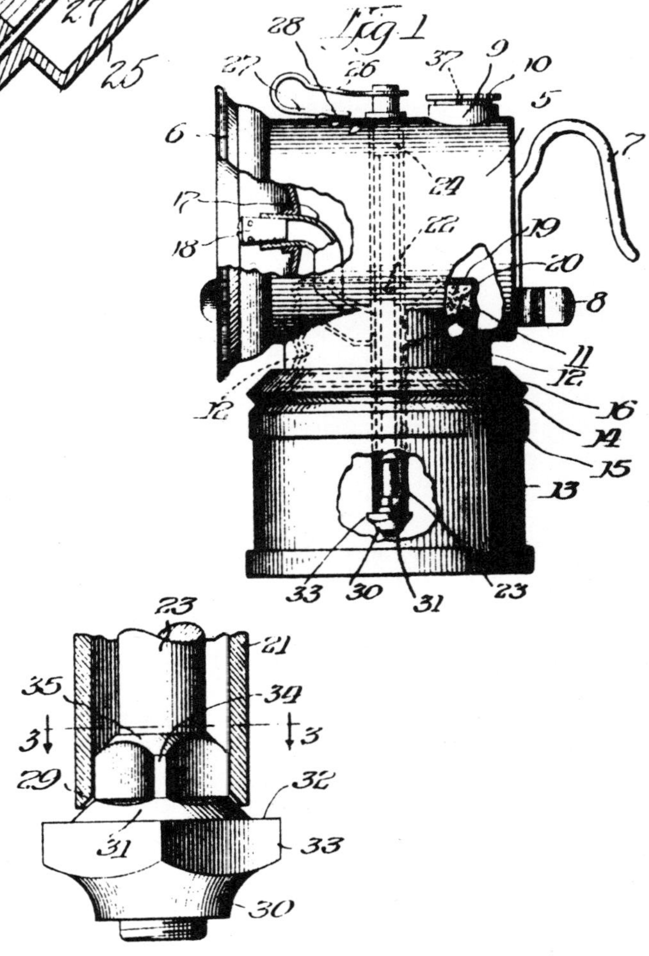

(Above) Diagram from Patent No. 1,407,141 dated February 21, 1922, to William J. Frisbie of Justrite Manufacturing Company. According to the patent specifications as described by Frisbie, "In prior valve structures of this class, the valve head is of transverse circular cross section and when this form of head is turned it cannot accomplish breaking up or agitation or stirring of the accumulated carbide sludge, so that with such structure the proper flow of gas is soon interfered with. One of the important objects of my invention is to make the valve head of irregular form, that is, to make it polygonal, or provided with projections so that when the valve stem is turned, the head will cut and push aside and thoroughly stir up the sludge around the end of the water tube so as to open up channels for the flow of the discharged water to the unslaked carbide."

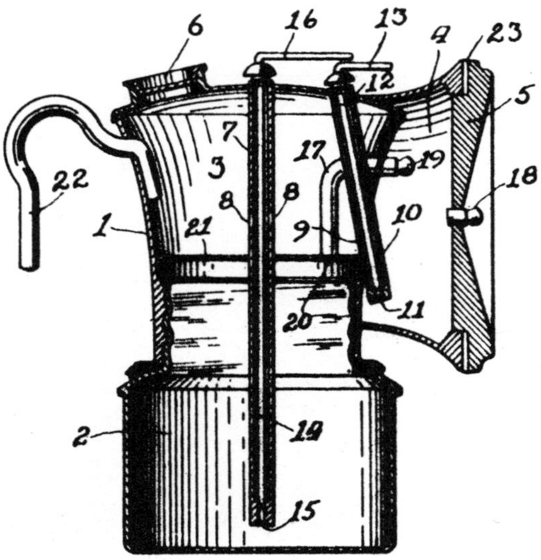

(At left) Diagram of Patent No. 1,106,062 dated August 4, 1914. Note the auxiliary carbide chamber located behind the reflector, 4.

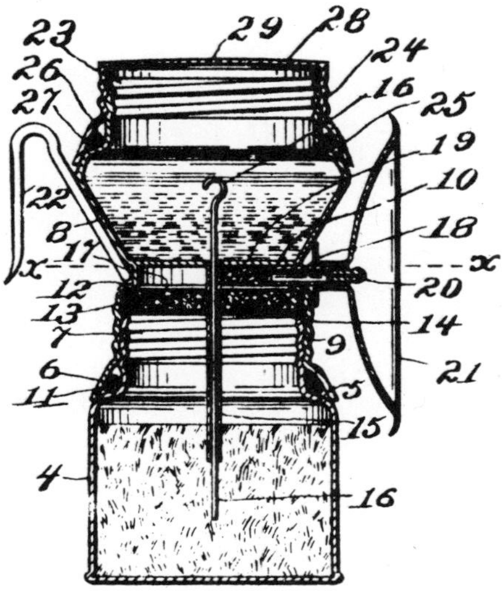

(Above) Togleson's Double Felt Filter, 1910. Note the gas space, 17. This was the predecessor of the gas spacer disc used later by Universal Lamp Co. and Justrite Manufacturing Co.

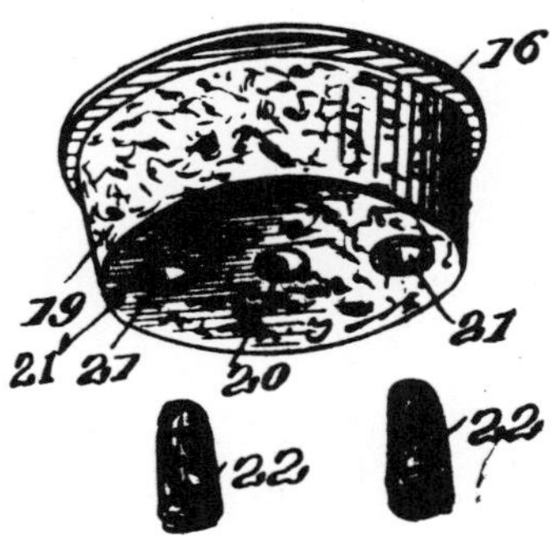

(Above) Acetylene gas lamp filter, Patented October 31, 1911, No. 1,007,374. Cork plug, 16, fits securely above carbide cup. Hole, 20, permits passage of water feed valve. Other holes, 21, are plugged by cotton, 22, acting as gas filters.

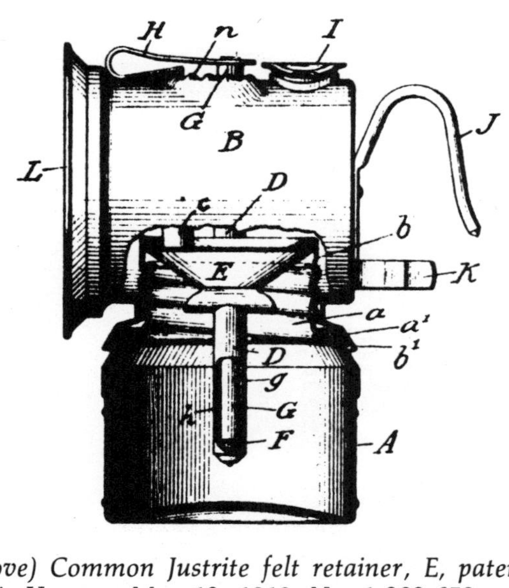

(Above) Common Justrite felt retainer, E, patented by Augie Hansen, May 13, 1919, No. 1,303,658.

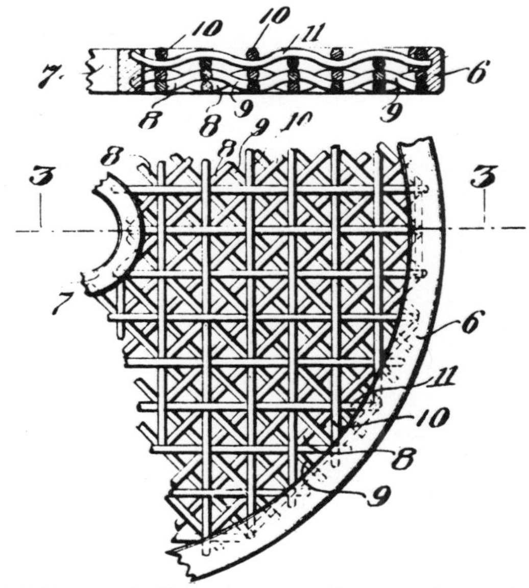

(Above) Wire mesh filtering screen for acetylene miner's lamp. Arthur Mitchell's Patent No. 1,282,913 dated October 29, 1918.

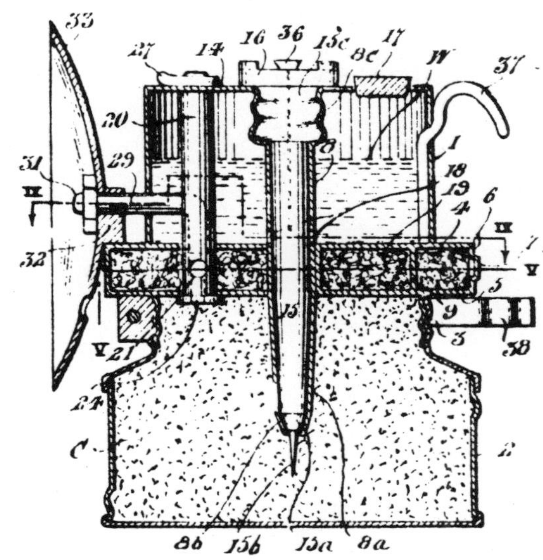

(Above) Menapace's spiral acetylene filter. Bisection of a carbide lamp showing filter, 19.

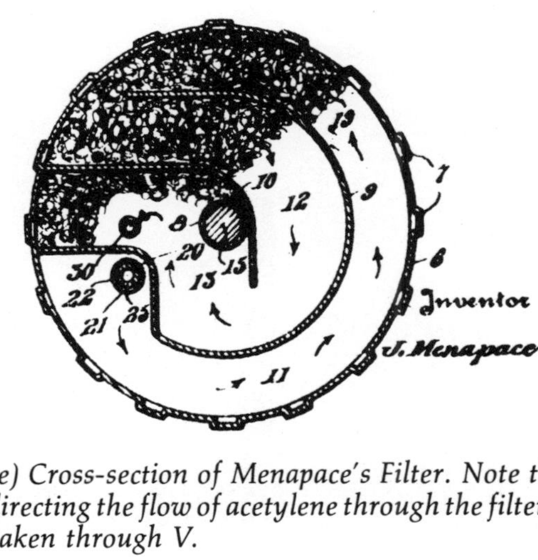

(Above) Cross-section of Menapace's Filter. Note the arrows directing the flow of acetylene through the filter. This view taken through V.

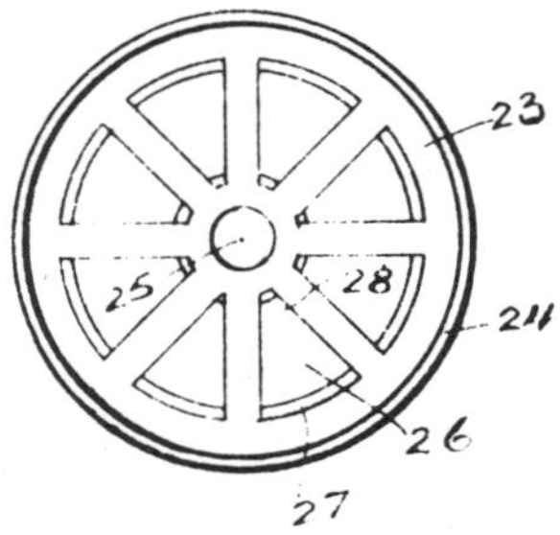

(Above) "Spider" or wheel filter retainer. Patented June 1, 1915, No. 1,141,475 to Arthur Mitchell.

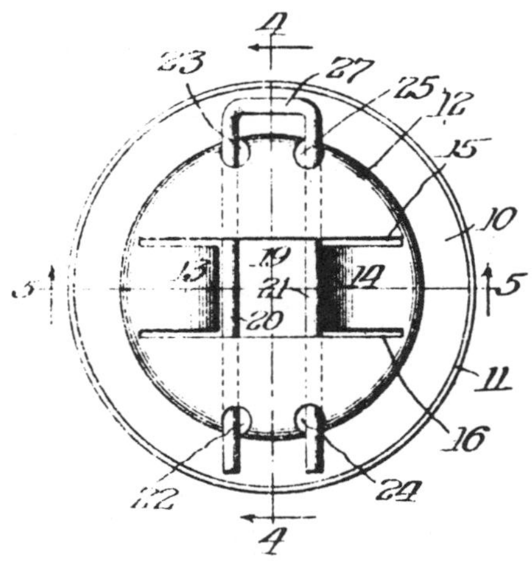

Larger retainer patented by William J. Frisbie, October 10, 1922, No. 1,431,447. Found in the eight-hour lamps manufactured by Justrite.

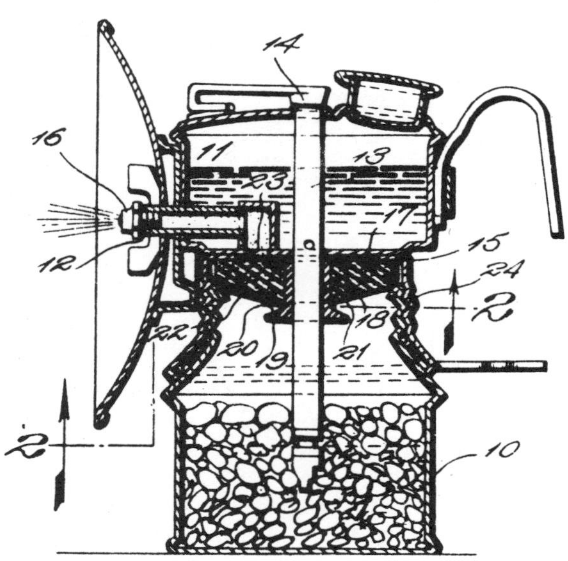

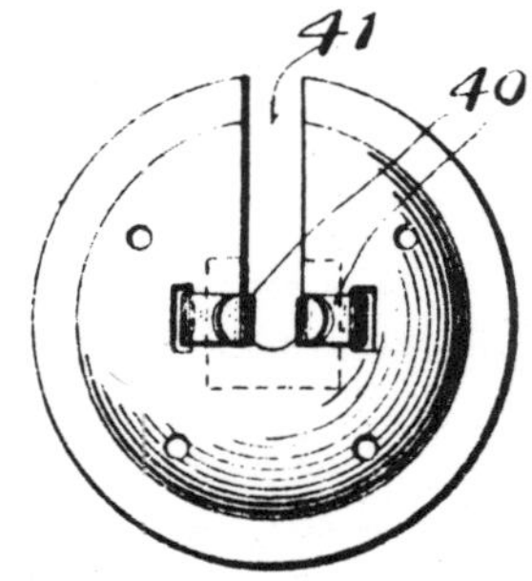

Holed retainer. One of just many varieties. Patented May 13, 1924, No. 1,493,802 to John M. Brock. Found in lamps manufactured by Dewar and Universal, among others.

An easier but related problem dealt with designing practical filter retainers or holders. Illustrated are a variety of retainers, all of which were used in miner's lamps.

Despite the various filters and the careful engineering they represented, the ability of the miners to "make-do" had more to do with providing an adequate gas filter than anything else. Whenever a lamp "acted-up," most miners simply reamed the burner tip. Failing in this, the next step would be to replace the felt particularly if it was wet or clogged with spent carbide. As the felts were inexpensive and easily obtained, replacement eventually became the most practical solution to a problem that was never fully resolved technically.

E. Reflectors

It was a simple matter of logic that the miner's lamp should possess a reflector. By concentrating and directing much more of the light than a mere open flame, a brighter, more efficient lamp could be designed.

A survey of early bicycle lamps shows cumbersome, glass-enclosed reflectors, impractical underground. As the lamps were considered for possible use in mining, a metal, reflective device, usually of polished brass, chrome plate, or nickel plate was tried as a reflector. Normally circular in shape and concave toward the burner tip in the middle, carbide lamps thus equipped were noted to be up to four times brighter than unequipped models.

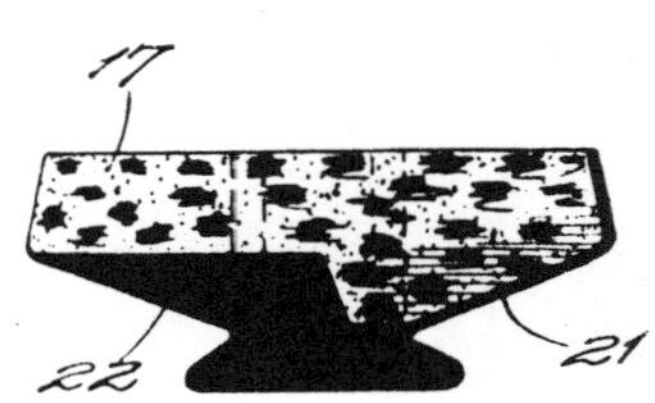

Impractical sponge rubber filter patented by Charles W. Schafer, Galena, Kansas, November 2, 1937, No. 2,098,167. Rarely used if ever.

Most of the early reflectors were simply fitted over the burner tube and tip and secured with a wing-nut or locking screw. However, there were some notable improvements advanced.

On May 7, 1912, Augie Hansen of the Justrite Manufacturing Company patented a lamp design with the reflector mounted flush into the water chamber. This was advertised as giving more stability to the reflector and its mounting. The model proved popular for Justrite and was manufactured for a number of years (Design patent No. 42,460).

One of the first experimental reflectors to deviate from the standard "wing-nut" design involved "a combined reflector and burner which may be formed from a single piece of material and will consequently be inexpensive. . . ." (Longstaff-Moore Patent)

One of the more bizarre reflectors, in fact, one of the more bizarre lamps was that illustrated in Wolfgang Gschwendner's patent of September 23, 1913. Not all of the reflectors were round!

Later that same year, Llewellyn M. Evans patented a reflector holder with a snapping, top cross-bar (6), as per the illustration. Evans was associated with Scranton Acetylene Lamp Company of Scranton, Pennsylvania, manufacturers of the "Scranto" brand miner's carbide lamp.

On January 6, 1914, Frederic Baldwin patented his blast-protected reflector for carbide miner's lamp. (See illustration and discussion under Shock Wave or Blast Protectors.) This gas-filled bag theoretically served the dual purpose of stabilizing the flame as well as cushioning the reflector from bumps and knocks. Other lamps employed various "spacers" from thin sheet metal and often proved inadequate.

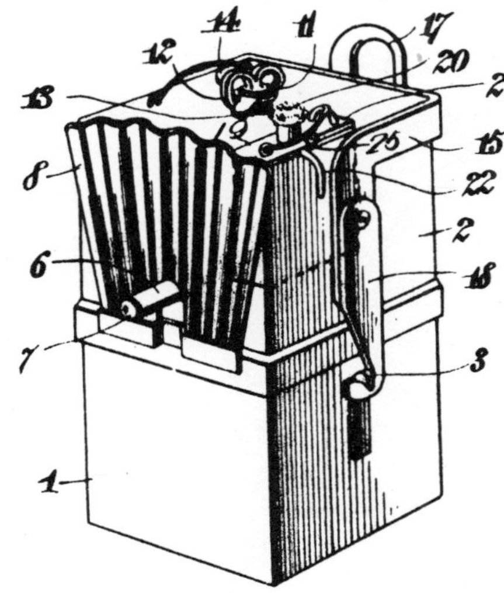

Gschwendner's Lamp with unusual, wave reflector. Note the small cotton wick 20, providing an auxiliary light source when recharging the water or carbide in the lamp. Gschwendner was from Cleaton, Kentucky, and it is doubtful if any of these strange lamps were ever produced in quantity. (See section on Unusual Lamps).

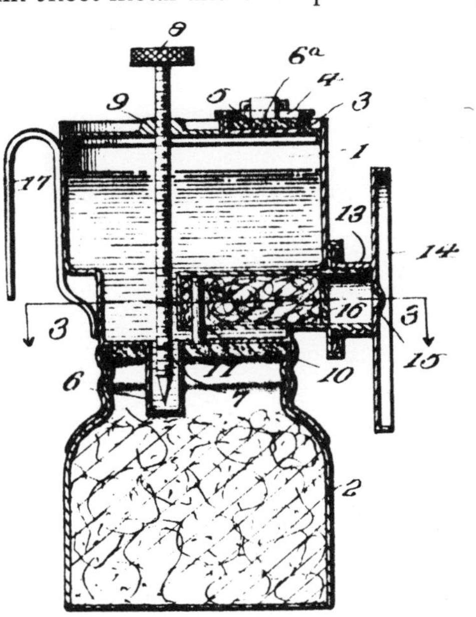

Thomas Longstaff and S. C. Moore's reflector patent. Note the design of the reflector 14, and burner tip 15. Patented February 18, 1913, No. 1,053,516.

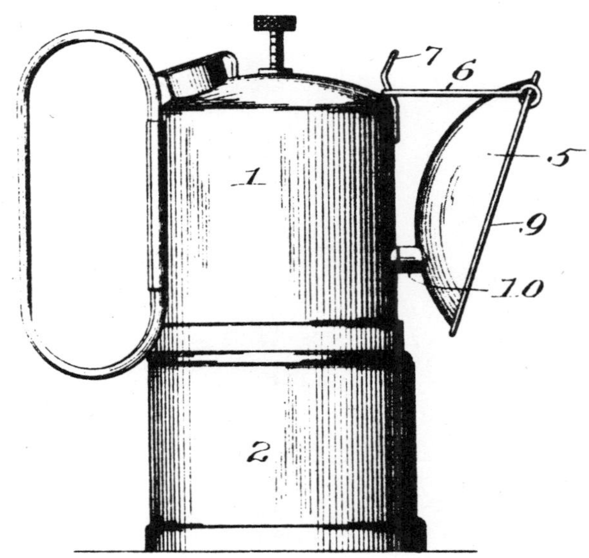

Diagram of Evan's Patent No. 1,081,899 dated December 16, 1913. Evans, a mine inspector, was associated with Scranton Acetylene Lamp Company, manufacturers of the Scranto lamp between 1913-1916.

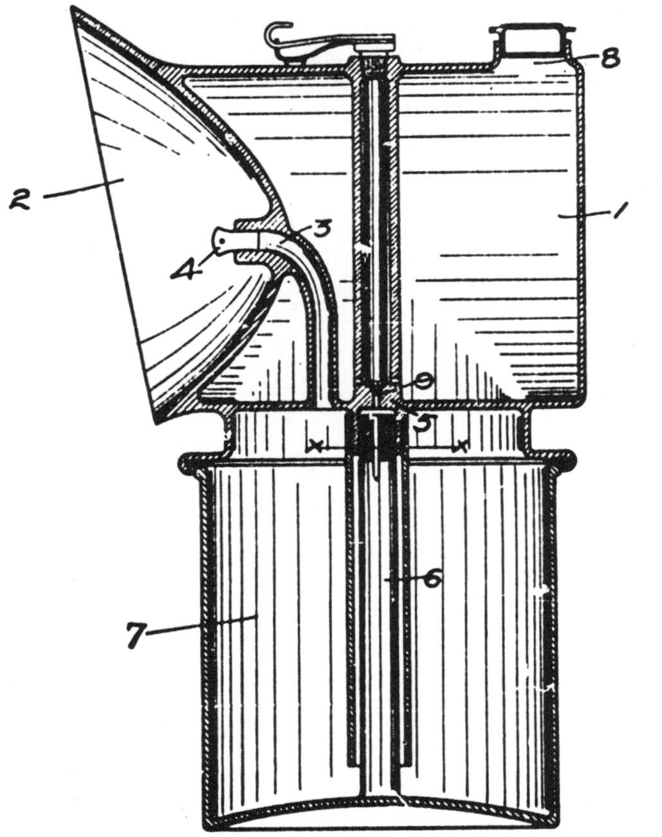

(Above) Diagram of Hansen's patent for cast lamp, patented June 26, 1917, No. 1,230,892. Note the reflector and front wall of water chamber are one. The "Giant" or "Little Giant" brand carbide lamp by Justrite Manufacturing Company was patterned after this patent.

Many miners simply improvised their own reflector braces, using washers, rubber gromets, or small coiled wire or springs to stabilize their reflectors. The variations were limited only by the miner's ingenuity.

There were several patents that tried to build the reflector "into" the lamp. Augie Hansen of Justrite patented a seamless lamp of cast metal that included a cast reflector sharing a back with the water chamber. This was the Justrite "Giant" or "Little Giant" brand carbide lamp.

Wilbur A. Cochrane of the Dewar Manufacturing Company patented a cast one-piece reflector designed to fit directly to the lamp body. Well-engineered and designed of light-weight aluminum, this reflector for some reason never became popular.

The three most popular miner's carbide lamps, Justrite, Auto Lite, and Guy's Dropper, patented separate reflector holders, each stamped of brass to fit the exterior of the lamp body. Probably functioning equally well, they were used for many years without variation. Only Justrite varied the design, using a dish-like, brass attachment for their holder in their "Streamlined" cap lamp introduced in the late 1930's.

There were a few lamps that had no reflectors. These had a gas tube exiting from the lamp body either on top or on the side. Generally classed as acetylene generators, these lamps simply did not provide enough light to compete with reflector lamps though they saw some use in coal mines.

Most lamps, however, had a reflector. The early bicycle lamps enclosed the reflector in glass, shielding the delicate flame from moving air. Bulky and awkward, they were not well suited for the miner's use. For this purpose, flat or concave reflectors were intro-

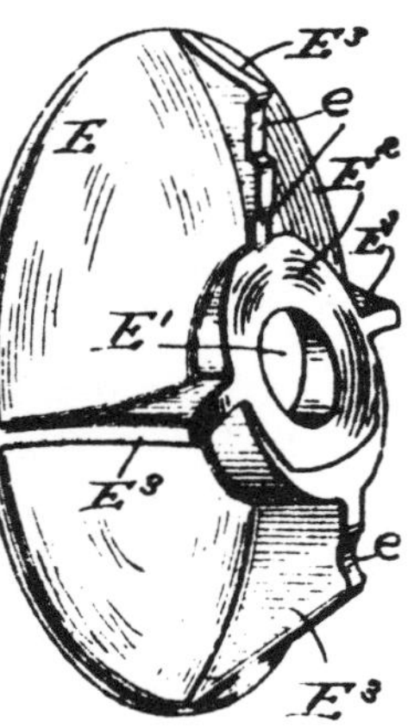

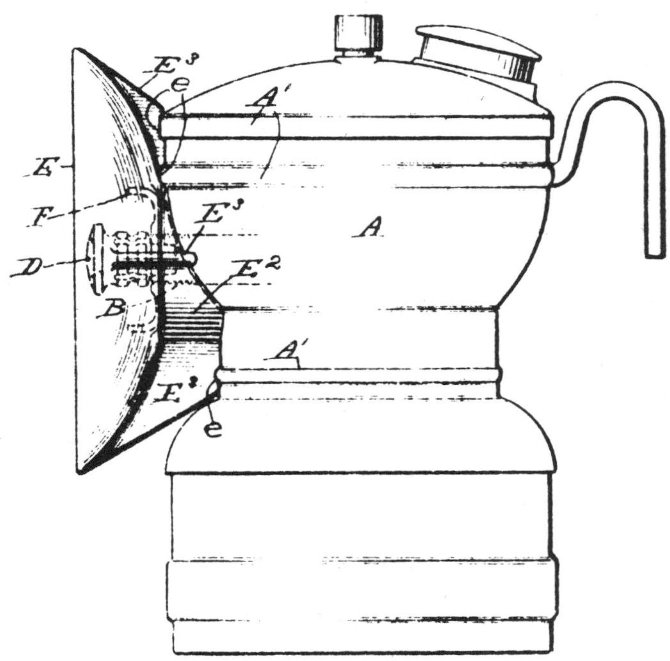

(Above) Views of Cochrane's one-piece aluminum reflector, attached and (to right), separate. Patented March 25, 1924, No. 1,488,193. Dewar Manufacturing Co., New York.

duced, usually fabricated from polished brass or plated with chromium or nickel.

As the concave design grew in favor, various sized reflectors were offered. From the common and popular 2½ inch model, there were also the 3, 4, and 7 inch reflectors, the latter appealing primarily to hunters, fishermen, and outdoorsmen. Some miners, wishing to do double duty with their lamps, simply replaced their smaller reflectors with old car headlight reflectors when they wanted to go frog gigging or coon hunting. Also popular among hunters and fishermen were the belt or hip mounted acetylene generators connected by a rubber tube to a hat mounted reflector. The Justrite and Searchlight brands were particularly popular in the twenties. (Note the illustration in the advertisements.)

As noted previously, many coal miners blackened their reflectors in an effort to reduce the glare. Other

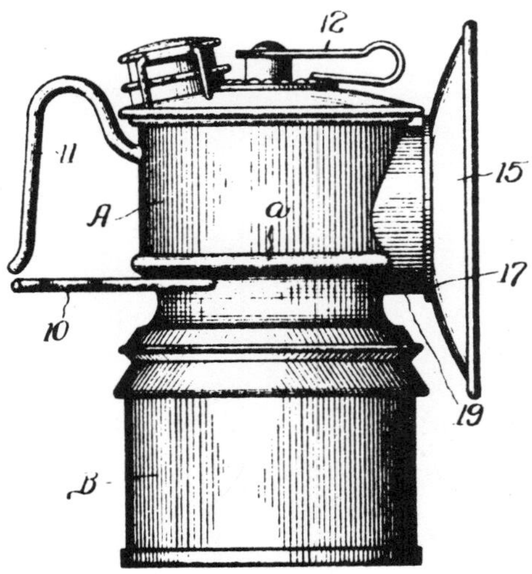

(Above) Side view of Justrite cap lamp. Reflector brace is designated as 19. Patented August 1, 1922, No. 1,424,464 to William J. Frisbie. (Below) Side view of Auto-Lite cap lamp. Reflector brace 19, is keyed to reflector by slot 23, and finger 22. Patented October 28, 1924, No. 1,513,068 to Adolph C. Recker.

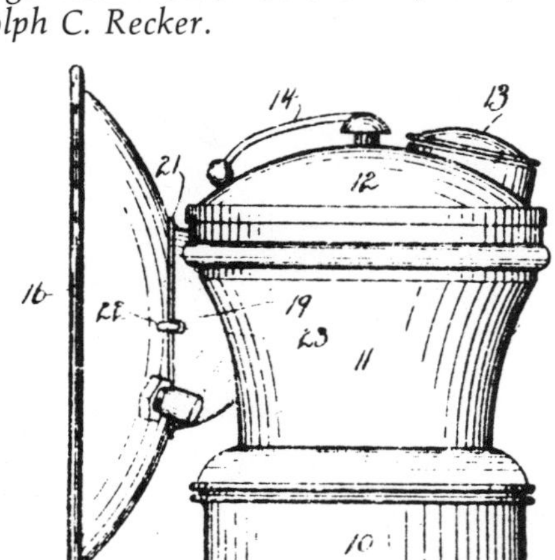

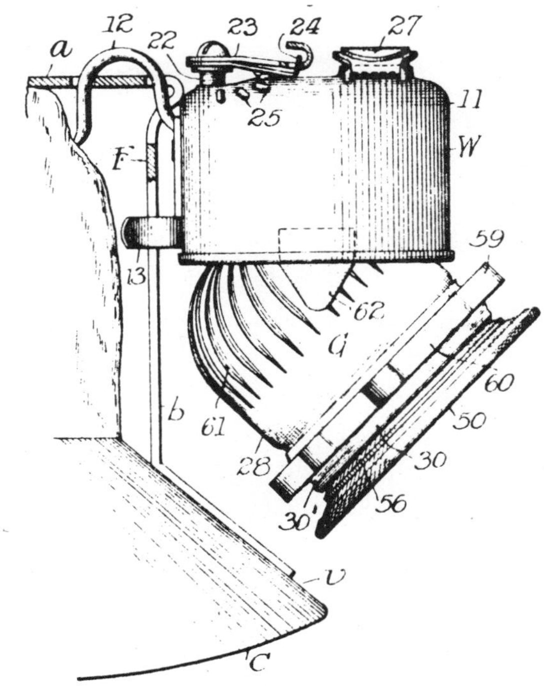

(Above) Another of Augie Hansen's elaborate lamps. Taken from Patent No. 1,165,969 dated December 28, 1915. Typical of Hansen's sophisticated engineering are the delicate water feed mechanism and swival device of this lamp. Again, this lamp was too fragile in design for use underground.

(Below) View of Patent No. 1,490,175 issued April 15, 1924, to Henry Luke of Neffs, Ohio. This was an attempt to mount a Railroad Hand Lantern to a miner's cap lamp.

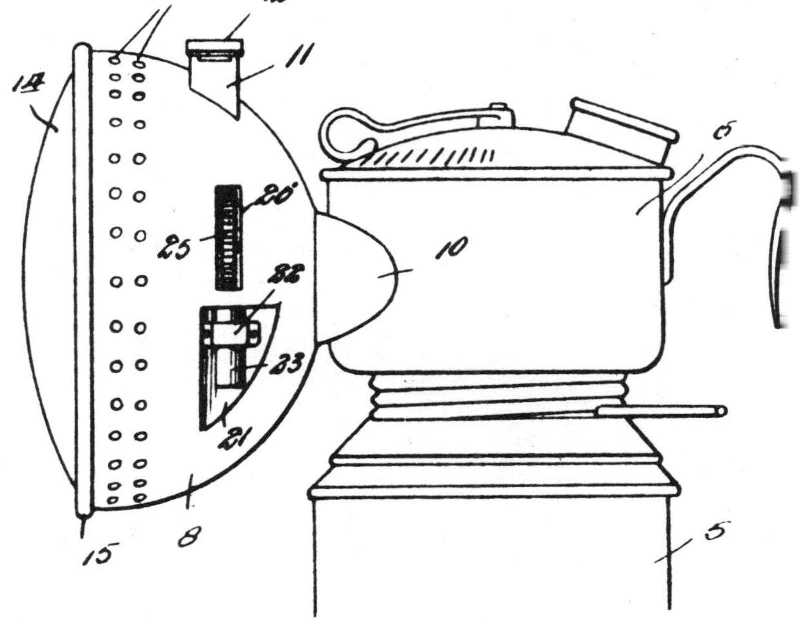

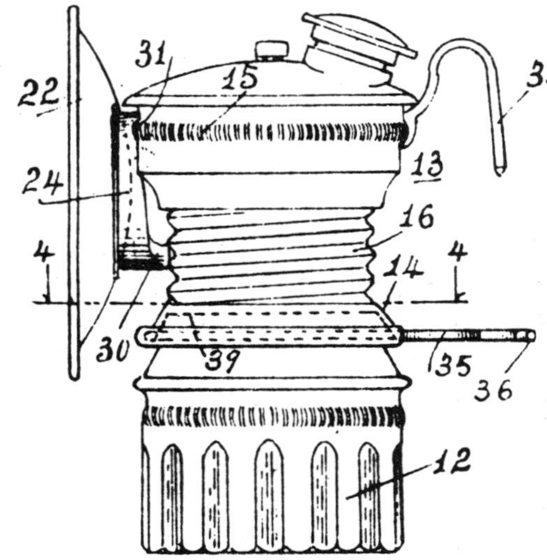

(Above) Side view of Guy's Dropper cap lamp. Note the reflector brace 24. Patented February 10, 1924, No. 1,525,579 to George W. Frost and Ernest O. Zacharias.

miners, particularly in metal mines, wanted more illumination and welcomed the introduction of the parabolic reflector. This reflector used the simple geometric principle of the parabola to concentrate the flame's light rays more efficiently, thereby permitting illumination at greater distances.

The opposite of this was to sacrifice deep penetration for a more diffuse, yet brighter light closer to the miner. This was the purpose of the "bullseye" reflector. By interrupting the smooth reflector with concentric rings, the light concentration and the direction of reflection would be scattered, spreading the soft glow around a wider work area for the miner.

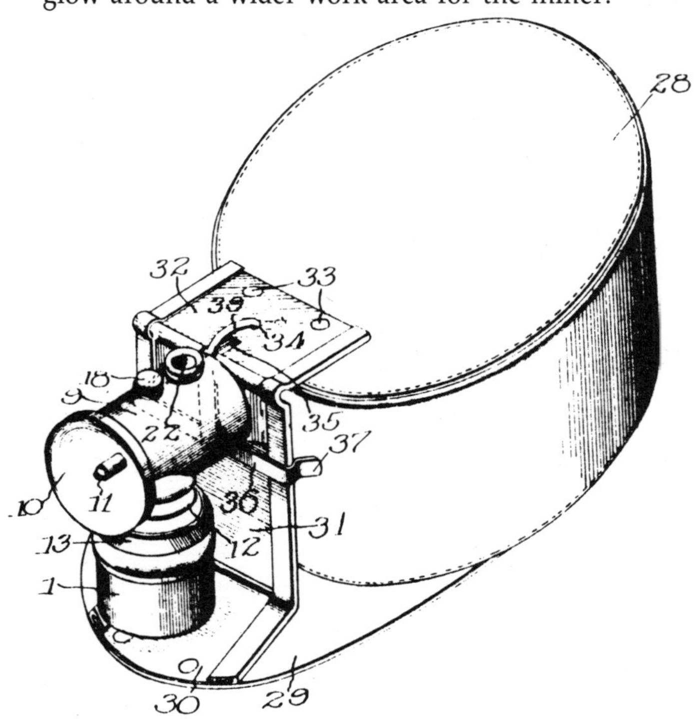

F. Lamp attachments, handles, cap hooks

With the development of acetylene mine lamps, most designs could be placed into one of two categories: Hand lamps or cap lamps.

Hand lamps, while widely adopted in Europe, were slow to "catch on" in the United States. Mainly used by the mine inspector or superintendent, these lamps had to be constantly moved by the miner while working, reducing his efficiency in the mine. Many utilized the hook and bail as exemplified by the "Acme" and "Big Boy" lamps. Some few incorporated a candleholder around the carbide chamber; see the "Justrite No. 77." And most employed simple handles for the miner to hold; see the "Dewar No. 160" for example.

In America, the cap lamp gained more favor than the hand lamp, permitting a miner to have light wherever he turned his head to work. Augie Hansen patented one of the early ideas for cap lamp attachments in 1915.

William J. Frisbie, working for the Justrite Manufacturing Company, patented another popular device for attaching lamps to caps. Employing a specifically bent wire interlocked in the threads of the carbide cup, the patent stabilized the pin hook attachment then in wide use.

A few years later, John B. Marco patented the blade attachment for carbide lamps. This patent, showing a drawing of the "Guy's Dropper" lamp, was soon applied to the "Autolite" and "Justrite" lamps, becoming very popular in later years with cavers. (See Letters Patent No. 1,554,451 dated September 22, 1925 to John B. Marco.)

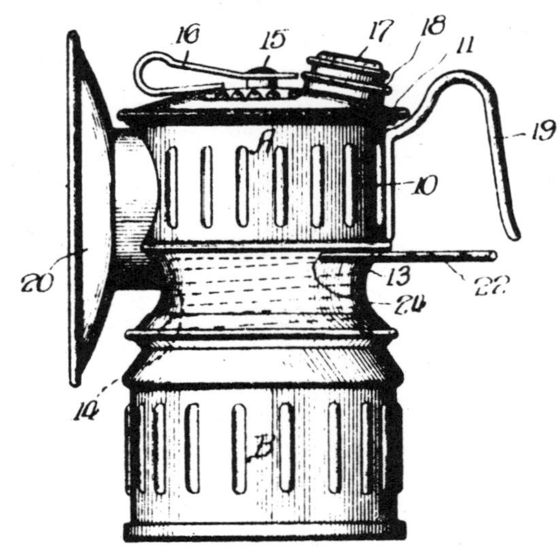

(Above) Frisbie's patent of August 1, 1922, No. 1,424,463. Note wire 22, passes into slot 24, and is held fast by the threads of the carbide cup and generator. (At left) Hansen's patent of November 23, 1915. This design was widely used by miners' cap makers for a number of years. Letters Patent No. 1,161,209

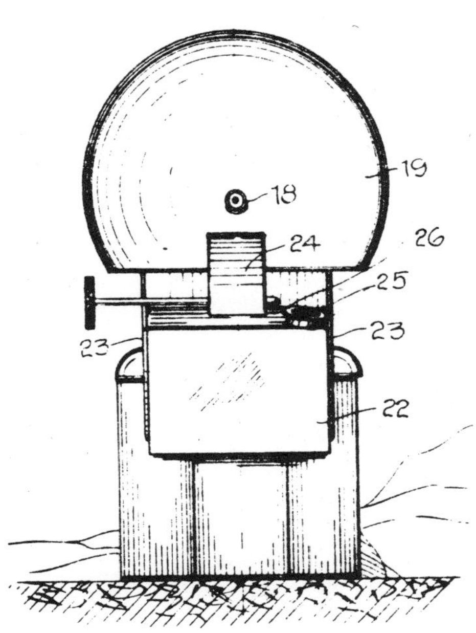

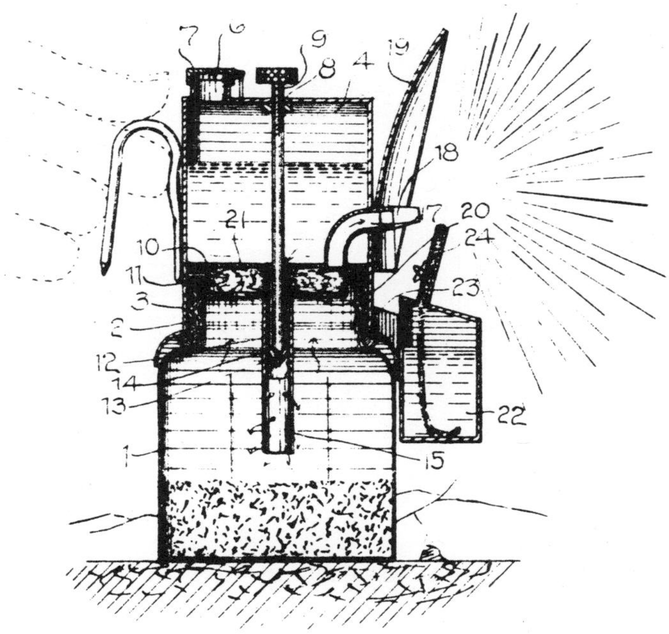

(Above and above right) Diagram of William F. Hudgins patent dated November 25, 1913, No. 1,079,347. This lamp was an attempt to provide the miner with a light source while refilling his carbide. Basically, the lamp included a small oil-wick lamp suspended from the water tank. Just before extinguishing his weakening carbide flame, the miner would crank forward the wick 24, and ignite it, providing himself with additional lighting during his carbide change.

Fig. 1.

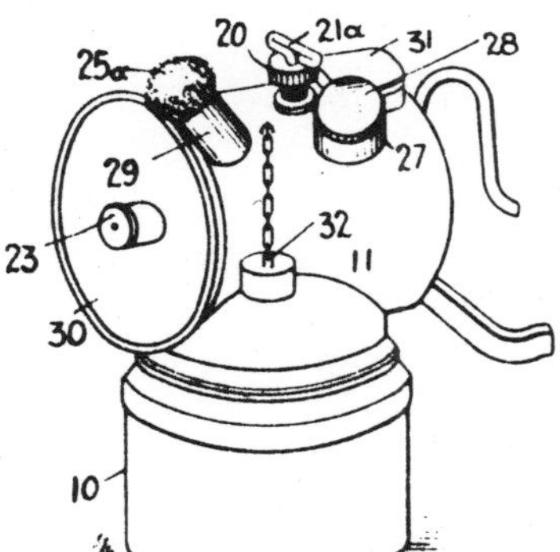

G. The Unusual

In the course of this study, a large number of acetylene mine lamp patents were examined. As has already been shown, many of these patents were either too intricate, elaborate, or fragile for practical, rigorous use underground. For the interest, study, and yes, amusement of the reader, the following pages are included as a small representation of the more unusual, even bizarre ideas that were patented.

Another similar attempt to provide an auxiliary wick lighting system on carbide lamps was patented by Michael Cotton, August 26, 1913. In a further attempt to solve the problem of supplemental lighting for a miner during a change of carbide, John Courtney invented an invertable lamp. Variations on this theme of supplemental lighting continued, almost to the point of absurdity. Note the intricacies of Paul Werner Lohmann's lamp which included two tips,

Fig. 2.

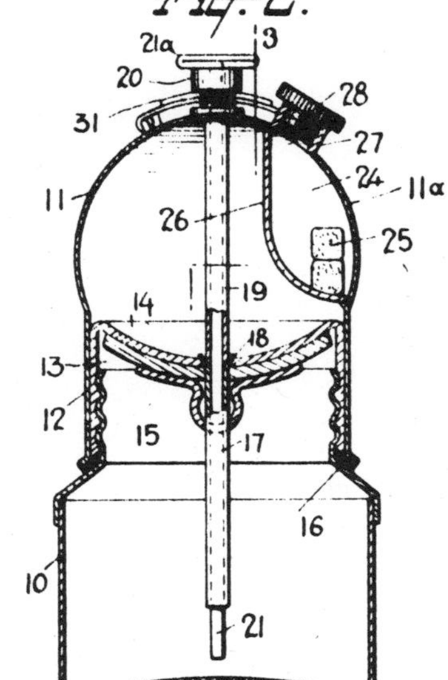

Fig. 3.

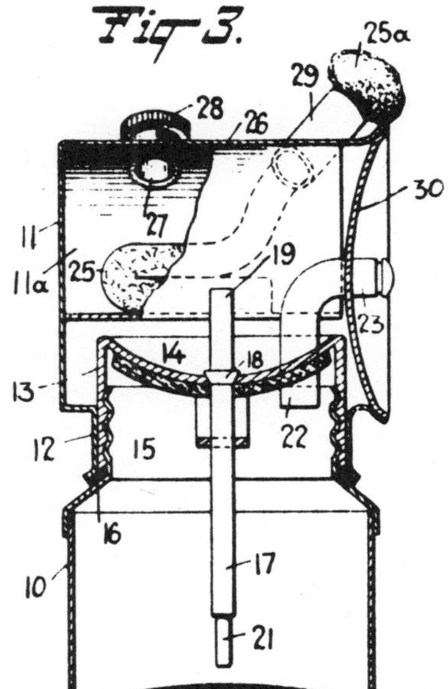

(Three figures at right) Diagram of Patent No. 1,071,550 to Michael Cotton. Cotton's idea incorporated an oil-wick 25, emerging from the top of the water chamber 11. A separate oil chamber 24 was built inside the water chamber. The Ever-Ready lamp, manufactured by the Charles Hoppe Company of Cincinnati, Ohio, was probably the only lamp to be provided with an oil lamp as an auxilliary recharging light.

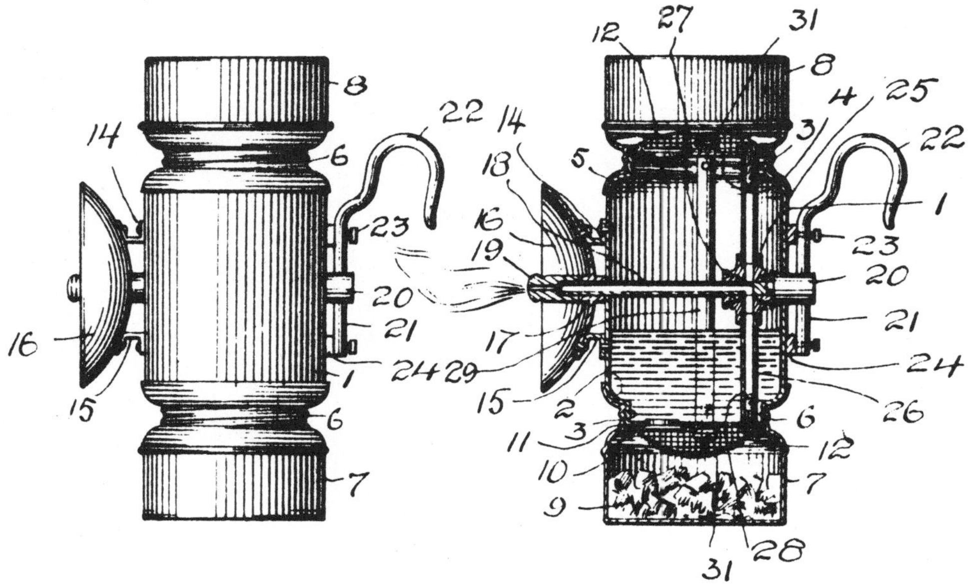

(At left) Diagram of Patent No. 1,063,419 to John Courtney dated June 3, 1913. This lamp in essence had two carbide chambers 7 and 8, and two separate water dropper systems 28 and 29. The lamp worked in the usual manner, water dripping down on carbide and generating acetylene gas. However, this lamp could be turned upside-down, and the water would flow through the corresponding dropper system and into the underneath carbide chamber. It was never stated what might happen if the lamp was positioned with both carbide chambers horizontal.

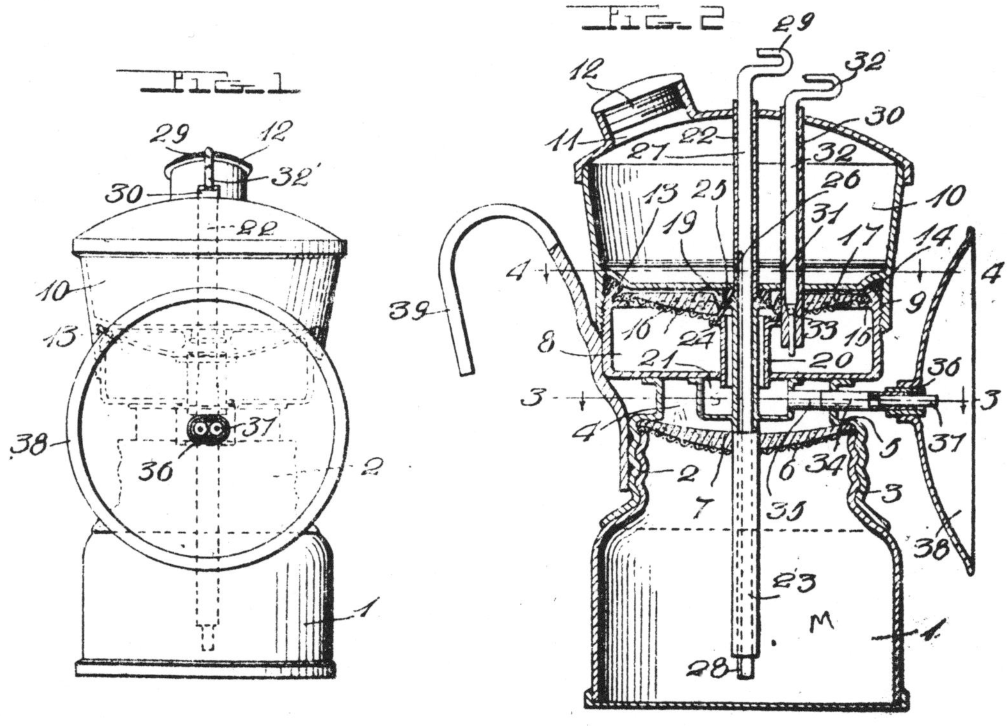

(At left) Diagram of Patent No. 1,078,865 to Paul Werner Lohmann dated November 18, 1913. This design incorporated two tips 36 and 37, two water control valves 29 & 32, to their respective carbide generating cups 1 and 8.

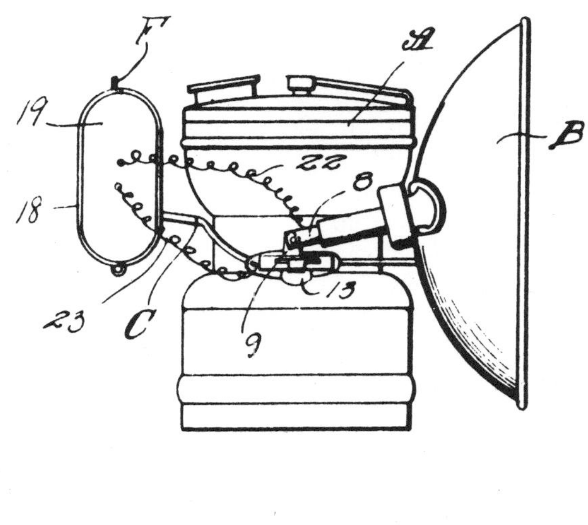

(At right) Diagram of Tony Bogi's Patent No. 1,358,278 dated November 9, 1929. A battery 19, powered a small bulb that was held and supported in a hole in the reflector.

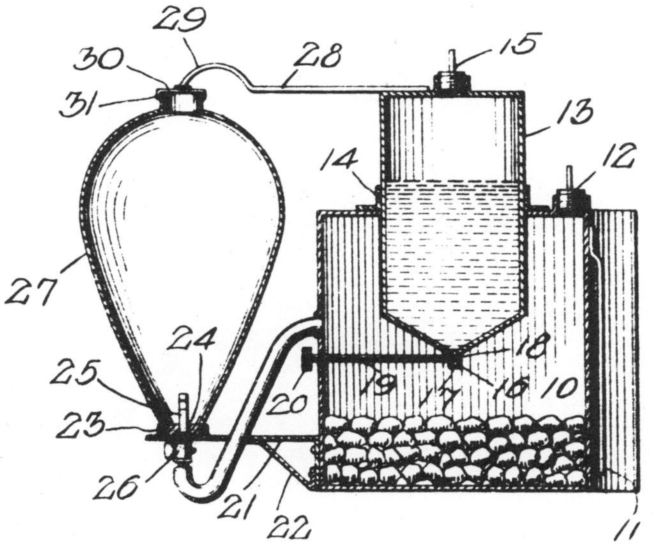

(Above) Patent No. 976,785 to Martin J. Curtis dated November 22, 1910. This most unusual design uses a horizontal water dropper device 16-18. Notice the fragile "burner globe" of glass 27. According to the inventor, this lamp was designed for wear on the cap.

two separate generating carbide chambers, two separate water feed systems and yet, only one water supply.

Other interesting and odd designs were advanced and patented. Take the nearly double lamp of Samuel T. Scoby. It possessed separate carbide chambers, water feed mechanisms, and burners while sharing a common water chamber and reflector.

Still another way to provide auxiliary lighting for carbide lamps was to employ electricity. Tony Bogi was one of the first inventors to patent an electrified carbide lamp.

As has been seen in previous sections, some of the early lamps were rather radical in design and form. Take for example, the lamp portrayed in Martin J. Curtis' patent.

Fig. 1.

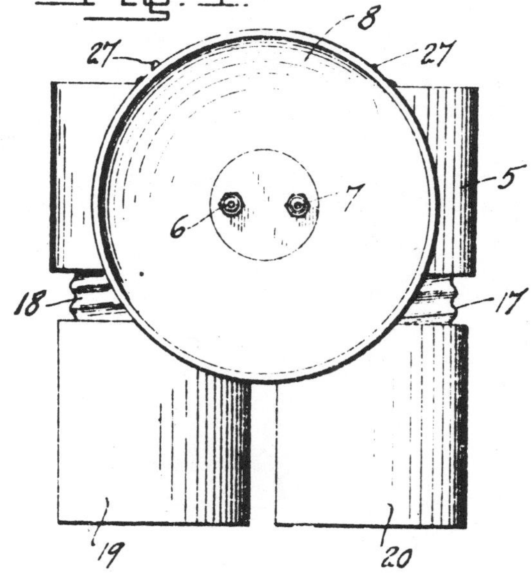

Fig. 2.

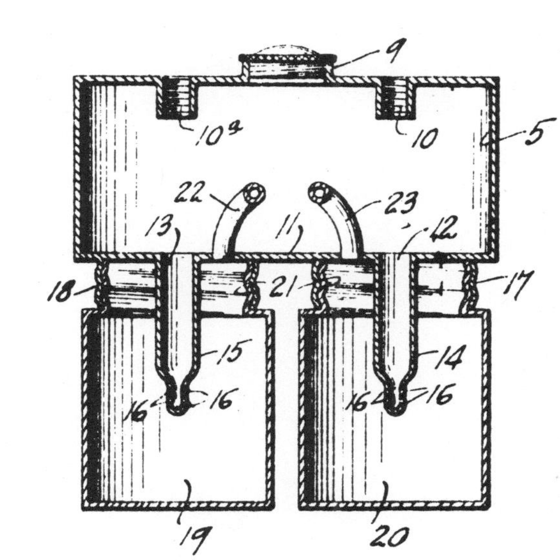

Fig. 3.

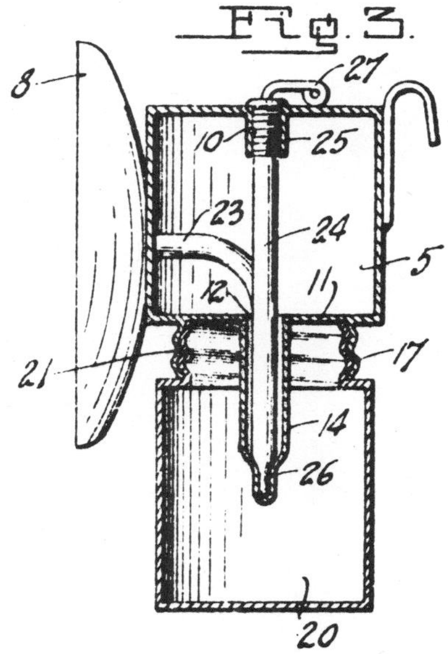

(Four illustrations at right) Diagram of Patent No. 1,567,682 to Samuel T. Scoby dated December 29, 1925. Interested readers should compare this sketch to the Maumee Duplex lamp manufactured by the Maumee Manufacturing Company of Pittsburgh, Pennsylvania, circa 1914.

Fig. 4.

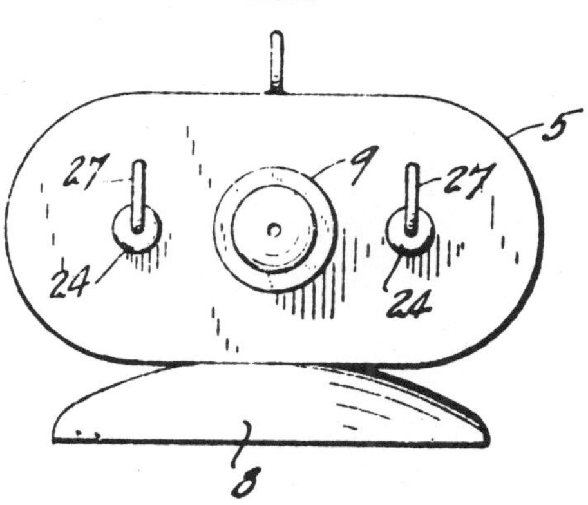

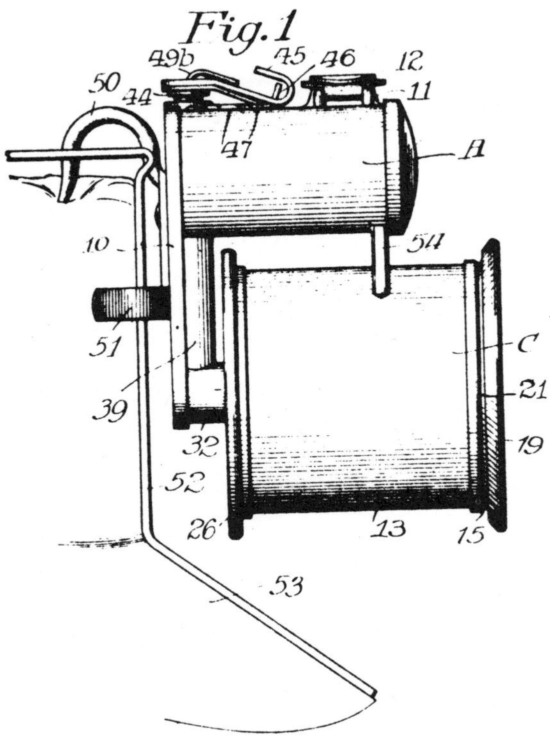

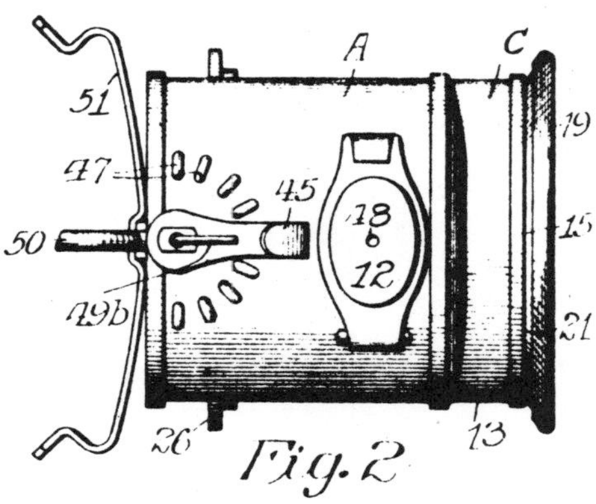

(Left and above) Diagram of Patent No. 1,180,492 to Augie Hansen dated April 25, 1916. "A" indicates the water chamber, 39 marks the water valve, and "C" is the carbide chamber. Figure 2 is a superior view of the lamp.

(At right) Diagram of Patent No. 1,198,608 to Fred E. Wilkison and George J. Mitchell dated September 19, 1916, the "carbide pocket lamp."

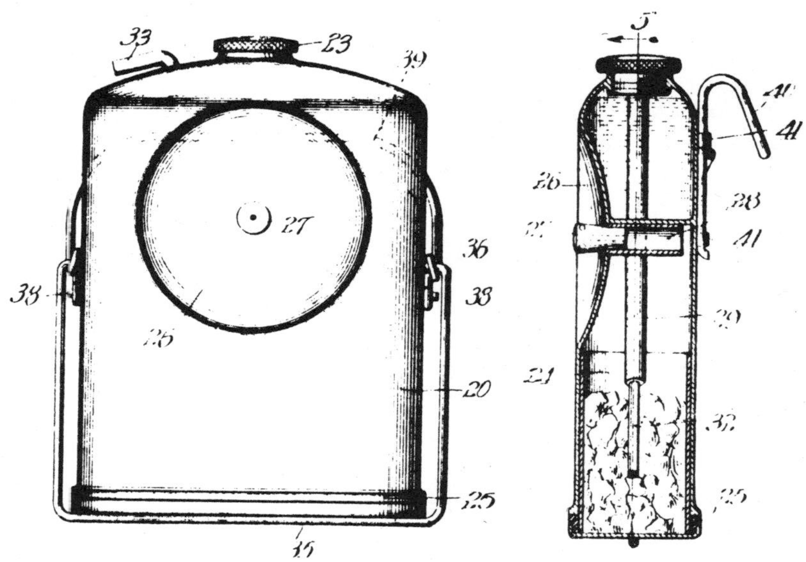

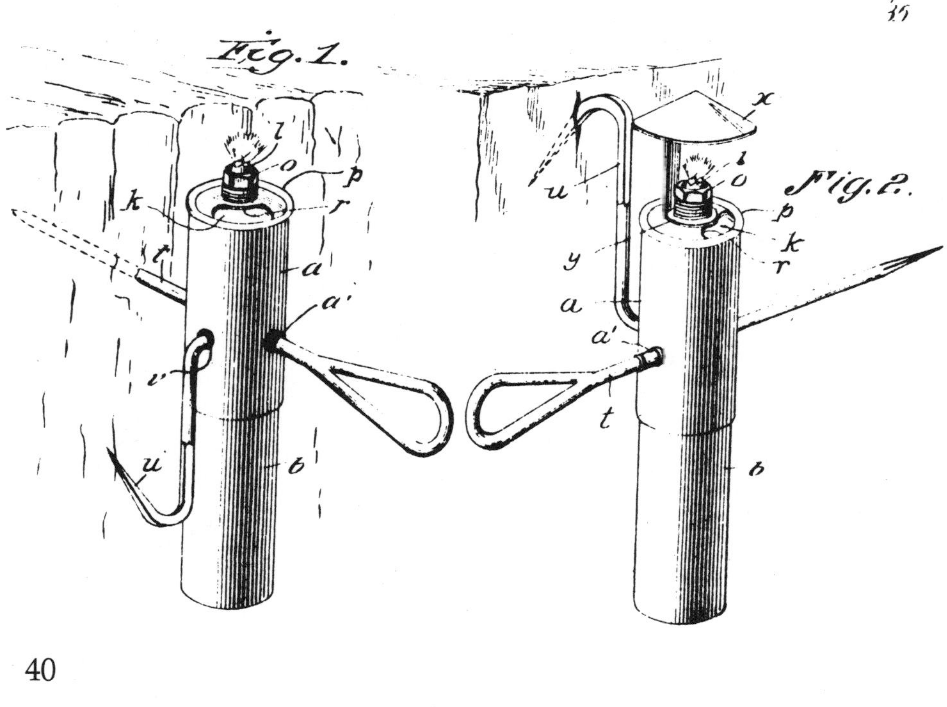

(Left) Patent No. 1,420,133 to Carl H. Neusiis dated June 20, 1922. From the diagram, it is obvious that this lamp was designed to appeal to the candle-loving miner. The lamp might be something of a cross between the "carbide candle" and a Varney candle holder from earlier years. About the only thing close to this that realized production was the Justrite models Nos. 101, 103, and 108.

Fig. 1.

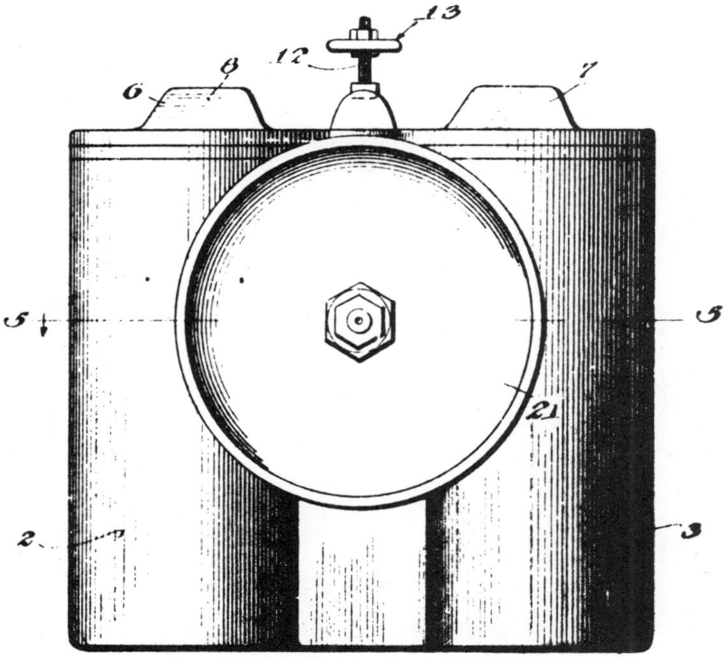

Fig. 2.

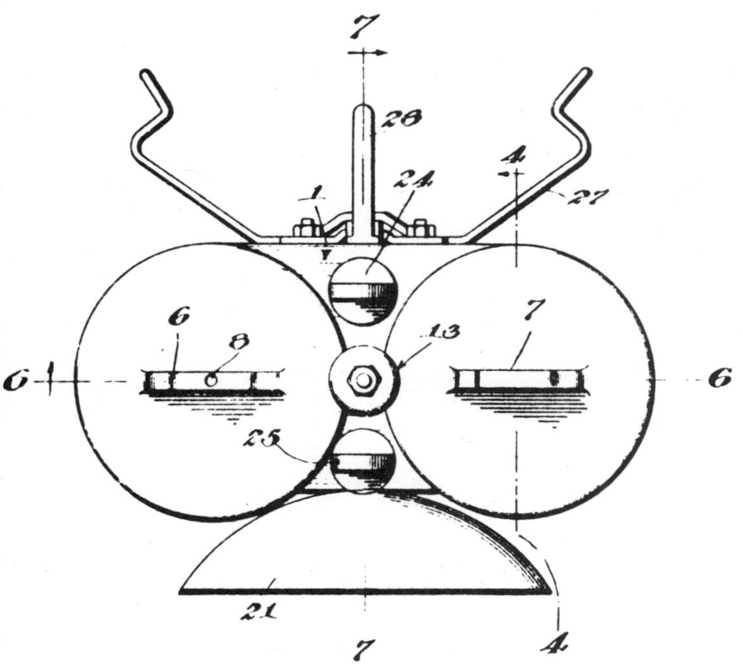

(Above) Diagram of Patent No. 1,693,463 to Walter Prockter dated November 27, 1928. With carbide located 2, and water held in 3, water control valve 12 released water through a small valve at the bottom of the lamp. Clogging of the mechanism by spent carbide was probably a definite drawback to this lamp. There were perhaps several models of this lamp fabricated with several prominent features incorporated in the design.

(Above) Diagram of Patent No. 2,214,546 to Domingo Anglada dated September 10, 1940. This lamp was designed to prevent flame sputtering by removing the moisture in acetylene with a water condenser 11, cooling the hot acetylene with water remaining in the water chamber 1. This unusual design resembles Wolf lamp No. 762, advertised in the mid 1920's.

Augie Hansen, that prolific inventor for Justrite, patented an unusual cap lamp in 1916. As with most of these odd designs, it was never manufactured.

It was probably inevitable that someone would patent a pocket acetylene mine lamp. Small, convenient, this lamp was probably never manufactured.

As the carbide mine lamp became more popular among miners, there were still some who held to the older ways of doing things. They preferred candles and that was that! On file at the United States Patent and Trademark Office are several "hybrid lamps" employing acetylene as the illuminant but retaining various nostalgic reminders of earlier times. Note the patent by Carl Neusiis on page 40.

There were other oddities. Walter Prockter designed a lamp that did not use gravity in the water feed mechanism.

Among the more interesting and possibly practical innovations that these odd patents produced was that incorporated into the lamp of Domingo Anglada of the Wolf Safety Lamp Company of America in New York City. Intended as a railroad lamp with an extension to "illuminate distant areas which are to be inspected," this lamp is seldom seen today. No doubt this lamp might be of use to cavers checking high, inaccessible leads deep underground.

No discussion on unusual, odd, or impractical lamps could be entirely accurate without a mention of the latest carbide lamp design from Justrite Manufacturing Company in Des Plaines, Illinois. With the advance of modern science, it is not surprising that a plastic lamp would emerge. Justrite holds the dubious distinction of introducing the first plastic carbide lamp in the United States. Without further comment, interested parties are referred to Letters Patent No. 3,735,120 of May 223, 1973 and No. 3,953,723 of April 27, 1976, both to Frank S. Flider of Justrite Manufacturing Company.

"Down in a silver mine . . . before carbide lamps."
—Courtesy Library of Congress

4

Carbide Lamps and Mining

By the beginning of the twentieth century, mine illumination had reached the threshold of technological improvement and change. A look at the "state of the art" showed miners relying chiefly on either safety lamps, oil lamps, or candles.

Oil lamps had been among the earliest mine lamps. Prints as early as 1556 show primitive metal miners carrying crude lamps consisting of wicks dipped in some kind of grease or oil.[1] Over the next several centuries, hanging oil lamps evolved from these "grease lamps" and were also employed underground. Cheap and convenient, these crude devices gave feeble light, but were the best the early miners had.

Candles were a natural outgrowth of these oil or grease lamps, but their open, exposed flame was easily extinguished in breezy passageways. This open flame was also being blamed for many of the disastrous fires and explosions occurring in gaseous mines. To venture into such places with open flame lighting was tantamount to playing Russian roulette.

This dangerous situation began to reverse itself with the introduction of the mine safety lamp, developed by Sir Humphrey Davy in 1812-16. The flame of this lamp was shielded by a wire gauze screen. When in the presence of methane or other flammable gases, the screen cooled the flame sufficiently to prevent the ignition of the gas. Miners found they could also check the height and color of the safety lamp flame and determine if the gas level in the passage was rising or dangerous. Refinements and improvements in the safety lamp by the end of the nineteenth century did much to increase mine safety.

Yet the safety lamp had its shortcomings. Its illumination power was not much greater than that of a candle. If a miner stopped to adjust the wick, exposing the flame to the open atmosphere of the mine, all safety features of the lamp were lost. Even with the safety lamp, it was apparent that the miner's ideal light source was still far from a reality.

The emergence of the oil-wick lamps around 1850 gave the industry a much brighter light than the safety lamp. This answered the need for increased illumination, but compromised this advantage with a smelly, sooty open flame. Questions concerning these lamps as fire hazards and detriments to miners' health raised few eyebrows. As it had been in the case of the early oil lamps, oil-wicks lamps were the brightest lamps available.

While candles, oil-wick, and mine safety lamps were the most common sources of mine lighting at the end of the nineteenth century, electricity was also being tested as a brighter, safer alternative. As early as 1880, the English had installed electric lights in a coal mine.[2] Portable electric lights were first reported used during the rescue at the Royal Pitt coal mine in

June, 1901, permitting the rescuers to work in gas-filled passages. By 1909, portable electric lights, though expensive, numbered nearly 2000 in English mines.[3] In the United States, the first electric mine lights were introduced in the metal mines at Ironwood, Michigan.[4] However, with electrical technology still in its infancy, miners turned to another new, more available alternative, carbide-acetylene lamps.

With calcium carbide being produced cheaply, acetylene could now be generated and studied at a low cost. Crude acetylene lighting devices were soon available, and the enthusiasts were quick to point out the advantages. Initially, mine lamp manufacturers were received with considerable skepticsm by the conservative, sometimes stubborn miners. The "Standard" acetylene mine lamp, first advertised in September 1902, converted few individuals and soon disappeared from the market.[5] The "Baldwin" lamp also had a small beginning, but persevered under several marketers; and by 1910, Baldwin and his manufacturer, the John Simmons Company, were the largest and best known acetylene mine lamp manufacturer in the United States.

In the years before 1910, the few[6] lamp manufacturers in existence struggled to establish a market with the mining companies. In an effort to persuade the miners, many of the advantages of the carbide lamp were enumerated, and comparison of the lamps with candles or oil-wick lamps was welcomed:

> Carbide lamps are used for the following reasons and purposes: To obtain better and stronger light; to improve the air of the mine by getting rid of smoke such as is made by oil lamps; to reduce the fire risk from snuffs and candles being carelessly left burning near timber or things that will take fire, or from discarded cotton wicks of oil lamps being thrown into the waste. . . .
>
> As compared with the ordinary miner's lamp or candle, a carbide lamp gives much more light. A carbide lamp is clean and the flame produces little smoke. The gases of combustion are principally carbon dioxide and steam (water vapor), and the acetylene is not injurious to health. The flame will not drop sparks and thus ignite explosives or anything that will burn as sparks from an oil lamp may do. With the better light the working place can be more easily examined and the dangerous roof conditions can be more readily detected. The general efficiency of the men is increased, because they are able to do more work with a better light. The cost of upkeep of carbide lamps is less than that of oil lamps or candles.[7]

Still, many owners and miners were reluctant to try the new lamps. The manufacturers advertised added improvements and offered to send free trial lamps to mine superintendents. The skeptics continued and often voiced their objections publicly.

> A short time ago, desiring to inspect a piece of work some distance inside the mine, and the mine boss not being

"The Standard" Acetylene Mine Lamp

Price $1.00

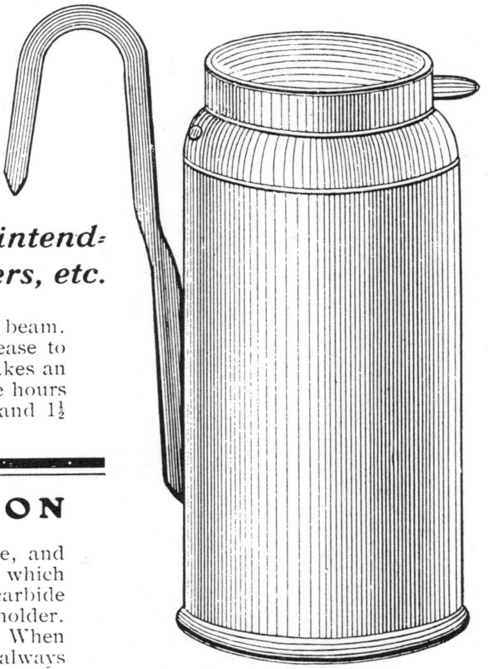

"The Standard" Acetylene Mine Lamp.

For Miners, Drivers, Trackmen, Foremen, Superintendents, Engineers, Surveyors, Inspectors, Timekeepers, etc.

Gives a brilliant light for all purposes, about 20 candle power in the beam. It is especially useful for examinations and surveying, there being no grease to soil the notes and maps. The flame is so small and clear that it makes an accurate point on which to focus the instruments. Burns from two to three hours on a charge of 1¼ ounces of carbide; weight 5 ounces, height 3 inches, and 1½ inches in diameter.

CONSTRUCTION AND OPERATION

THE lamps are made of brass so as to be practically indestructible, and are composed of three parts; first, the bowl or body of the lamp, which contains the water supply; second, the gas chamber; third, the carbide holder. With each lamp there is an extra gas chamber and carbide holder. This allows the changing of charges and have a light all the time. When the one in use is about exhausted, take the extra one, which should always be ready, and wet the wick and light it from the one in use, put a new supply of water into the bowl and replace. This is a feature no other acetylene lamp has. These lamps, when the simple directions are followed, burn the same as an ordinary gas jet, without any more smoke or smell, and do not burn out the oxygen, nor produce harmful gases, to the same extent. At the present price of carbide this light can be produced much cheaper than with oil.

Calcium Carbide $Ca\ C_2$ from which acetylene gas is produced is made by mixing powdered coke and lime together, and submitting this mixture to the intense heat of the electric furnace; when fused and allowed to cool it becomes a hard black mass. This when broken up and crushed into various sizes is packed in cans and made ready for shipment and is commercial Calcium Carbide.

Calcium Carbide has a great chemical affinity for water, and when brought into contact with it, the oxygen of the water unites with the calcium of the compound, liberating the hydrogen and carbon, which combine to form acetylene gas $C_2\ H_2$ and it is this gas which is burned $Ca\ C_2 + H_2\ O = Ca\ O + C_2\ H_2$.

Speaking of acetylene gas, Dr. Grehant shows that when properly burned, no carbon monoxide can be detected. By way of comparison, taking an averaged sized room, which would be well lighted by 64 standard candles, he states that this amount of light obtained from various illuminants, would show the following results.

ILLUMINANT.	OXYGEN REMOVED FROM AIR.	PRODUCTS OF COMBUSTION.	
		WATER VAPOR.	CARBON DIOXIDE.
Sperm Candles	38.5 Cu. Ft.	26.2 Cu. Ft.	43.6 Cu. Ft.
Parraffin Oil	24.9 "	14.0 "	39.8 "
London Gas	26.1 "	29.4 "	19.2 "
Argand	23.0 "	25.6 "	17.0 "
Regenerative	10.6 "	8.3 "	5.2 "
Acetylene	5.0 "	2.0 "	4.0 "

This shows that acetylene gas takes much less oxygen from the air than oil lamps or candles; also that it gives off no deleterious gases. In workings ahead of air this feature is most valuable, and, because of the better air, men work with more vim and can remain longer at the work. Acetylene gas, moreover, will burn in air so foul that other illuminants are extinguished.

Mr. James W. Paul, Chief Mine Inspector of the State of West Virginia, says of this lamp:

"Its general adoption in all mines in which explosive gases are not found would very greatly benefit the health and safety of the underground employes."

Write for copy of Chief Mine Inspector Paul's letter of indorsement, descriptive circular, etc.

AGENTS WANTED

John S. Cummings & Company

TUNNELTON, WEST VIRGINIA

Baldwin Acetylene Mine Lamps

give a light of from 10 to 16 candle power, consuming 8 ounces of carbide in 10 hours at an average cost of about 2 cents.

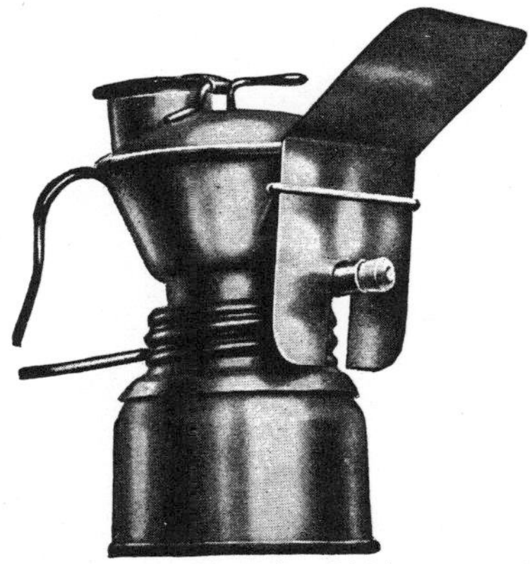

No. 33. Price $1.00.
Patented.

The Baldwin Acetylene Mine Lamp has received the unanimous endorsement of mine owners and miners throughout the country.

No. 31. Price $1.00.

Smokeless

Economical

Durable

No. 32. Price $1.00

It has proved itself much cheaper than any known method of mine lighting. It is reasonable in first cost, well constructed, light in weight, and so simple that a man of little intelligence can operate it.

No. 36. Price $2.00.

The "Baldwin" can be turned upside down, laid on its side or let fall, it will burn just the same.

Send for illustrated literature and get all the facts concerning the "Baldwin."

John Simmons Co.,

104 Centre Street, - New York.

Coal miners at the Zenith Mine, Butler, Pennsylvania.
—Courtesy National Archives

available, I went to the place alone. After satisfying myself as to the work, I decided to return by a short cut through some old workings, forgetting that "the shortest way home is the longest way "round." Part way through the old workings my lamp went out. I had a plentiful supply of carbide and water but the burner was stopped. The little piece of wire, soldered to a small piece of tin was missing. I had four matches and managed to get outside, but my faith in the acetylene lamp is badly shaken. . . . this point of view may not be a progressive one, but I feel there is yet much to do to the acetylene mine lamp before it becomes a practical success. . . .[8]

Yet there were champions of the carbide lamp in the mines, solid converts who were also ready and willing to publicly applaud it.

When first coming to this country I thought the oil lamp superior to the Mersaut safety lamp. After changing to the Baldwin acetylene lamp I felt at first that I should have to throw it away . . . but now I could not be induced to return to the use of oil lamps under any consideration whatever.

The carbide lamp is almost automatic in its action and I believe its success depends on this principle. I believe that the greatest difficulties in their use arise from improper handling.[9]

Slowly, with the help of testimonials like the above, miners (pardon the pun) saw the light. Having once used a carbide lamp on a trial basis and become familiar with the care and maintenance of the lamp, many users were convinced. Acceptance created demand and this demand was met by an increased number of different brands of lamps. By 1915 it was estimated that 300,00 acetylene lamps were in daily use underground in the United States.[10]

With acetylene lamps now being widely adopted for use in coal and metal mines, more than just subjective evidence was needed for its proper evaluation. How safe was acetylene and how much carbon dioxide did it produce? Quantitatively, how much brighter was an acetylene flame than the best oil-wick, safety lamp, or electric bulb?

To answer these and other questions, the United States Bureau of Mines began testing in 1911. It was found that at standard temperature and pressure, the burning of 1 cubic foot (cf) of acetylene required 2.5 cubic feet of oxygen or roughly 12.5 cubic feet of air. For the average carbide lamp using 12-14 grams of carbide per hour, this yielded 0.121-0.144 cubic feet of acetylene per hour. Using 0.36 cubic feet of oxygen or 1.8 cf of air, the reaction would release 0.288 cf of carbon dioxide.

With the lamp burning ten hours in a work area of 150 of (3 x 5 x 10 feet), 2.88 cf of carbon dioxide would be generated. When added to the 0.006 cf of carbon dioxide normally present in the atmosphere, 2.886 cf or 1.0% of the 150 cf of the working space would be

(Above)Miners headed underground for a work shift at Osceola Copper Mine Number 13.
—Courtesy Marquette County Historical Society

carbon dioxide. To put this into perspective, it was pointed out that a miner at moderate work uses some 3.2 cf of oxygen per hour, exhaling 2.56 cf of carbon dioxide, or more than 8.8 times the carbon dioxide than that given off by the carbide lamp.[11] Subsequent testing also failed to demonstrate any alleged acetylene toxicity provided adequate ventilation was maintained.

It had been long known that an oil burning lamp would go out when the oxygen content of the surrounding atmosphere dropped below 17-17½%. Testing revealed that the acetylene flame remained burning until the oxygen level fell to about 12½%. It was noted that

a man's life is endangered should the oxygen content fall much below 10%. At about 14% oxygen, however, the color of the acetylene flame changes markedly . . . the miner is given obvious and adequate warning . . .

. . . the warning of the oil-fed flame is given with so high an oxygen content that miners had learned to disregard it, and will frequently go into workings containing air in which their oil lamps will not burn.

Such is not the case with the acetylene lamp; its warnings are given so near to the danger point that men will have a wholesome respect for them . . .[12]

Candle power comparisons left little doubt of acetylene's superior illumination.

CANDLEPOWER OF VARIOUS TYPES OF PORTABLE MINERS' LAMPS[13]

LAMP	CANDLEPOWER	COST PER SHIFT, (cents)
Electric cap lamp	1.10	
Miner's open flame acetylene cap lamp	5.00	1.5
Miner's open flame oil cap lamp	1.50	2.4
Davey safety lamp	0.12	
Clanny safety lamp	0.35	
Wolf-type safety lamp	0.65	
Akroyd and Best safety lamp	1.10	
T. M. Chance acetylene safety lamp	3.80	

In light of these and other test results, the following conclusions were presented at the meeting of the American Institute of Mining Engineers in New York in February, 1917.[14]

1. The open flame oil lamp has outlived its general usefulness.
2. The use of the open acetylene lamp is growing in all non-gaseous mines because of its cheapness, the powerful light it gives, its reliability in the presence of the black damp, and its freedom from soot and noxious vapors.
3. The electric cap lamp is best adapted to use in gaseous mines
4. The flame safety lamp should be used in all gaseous mines irrespective of the use of electric cap lamps. . . . it must be used as a gauge of the safety of the mine atmosphere.
5. There is a new acetylene safety lamp that gives promise of being an admirable working lamp as well as an excellent fire damp and black damp detector.

By the time the United States entered World War One, carbide lamps had reached the height of their popularity. Even in coal mines where gas was not nearly as notorious as in Europe, acetylene lamps were widely used. Production figures for the popular "Guy's Dropper" lamp show that 720,000 lamps were manufactured in 1919 alone. For the next several years, it was the golden age of acetylene mine illumination.[15]

For most of the miners, changing to carbide from the sooty oil-wick lamps was nothing more than a

(Right) "Drilling for ore" at a mine in the Marquette Range, ca. 1916. —Courtesy Marquette County Historical Society. *(Below) Underground in the Springfield Coal Mine, ca. 1933.* —Neef Photo

(Above) In September 1922 rescuers working at the Argonaut Gold Mine, in California, which had exploded and collapsed, broke through to an isolated chamber deep in the mine and found this tragic message from miner Fezzel, one of the doomed men. With his failing carbide lamp, he was able to chronicle the deteriorating breathing conditions that eventually took his life. —Courtesy National Archives

change in routine. At the beginning of each eight hour shift, a miner would be given a full charge of carbide for his lamp and enough carbide in his storage flask for the duration of the shift. When the shift was over, he would carry out the used carbide and dump this into a screening pan, permitting all the unused carbide granules to be recovered and placed back into the carbide storage can for later use. Pity the careless miner who broke the rules and dumped his half-spent carbide in the wet passages of the mine and refilled his lamp with fresh carbide from his flask. Even if he ignored the pungent smell of the freshly generated acetylene and escaped the possible danger of an explosion, he still might run out of carbide before his shift ended and be plunged into darkness.

One of the first things miners noted about their new lamps concerned the average "burn time" of a charge of carbide. It soon became a rule of thumb that a lamp needed recharging when half of its advertised "burn time" had expired. Thus, a so called "four hour" lamp should be tended every two hours; eight hour or shift lamps should be recharged right after lunch. Miners, desiring as much light as possible in their dark, sometimes cramped quarters, often burned the lamps "wide open" and greatly reduced the duration of a single charge. Reaming the burner tips for increased gas flow permitted larger flames but shorter recharge times—and the more time a miner spent recharging or refilling his lamp, the less coal or ore he produced, and the lower his production tally at the end of the day.

Penny-pinching mine bosses, looking to slice overhead, sometimes made miners provide their own lamp supplies, including felts, tips, reflectors, flints, and occasionally even carbide. No wonder miners beginning a shift would often only light one lamp with the flint striker, then use this one lit lamp to ignite the remaining lamps, thus conserving flints and pennies.

The flasks in which most miners carried carbide

were of a wide and varied assortment. Many lamp companies also manufactured and distributed these containers; others were homemade by the lamp user. Some miners carried extra lamp bottoms already filled with carbide and covered with a screw cap top. This cap was sometimes punctured to permit the venting of acetylene generated by residual moisture in the lamp bottom. When their lamp needed recharging, they simply unscrewed the used bottom and replaced it with the new. Sometimes miners used racks for carrying multiple lamp bottoms into the mines, thus permitting easy transport of carbide into the mines and spent carbide out of the mine at the end of the shift.

As miners became accustomed to these lamps, they developed preferences and added certain modifications to increase their lamp's efficiency. Unlike their European counterparts, nearly all miners in the United States came to prefer cap lamps to hand lamps. It was a sure bet that any fellow with a hand lamp was probably green on the job.

One feature that the coal miners quickly changed was the bright, shiny surface on their reflectors. Originally chrome, nickel, or polished brass, the reflectors were immediately sooted and blackened by coal miners. The explanation given was that coal mines were much "darker" by nature than ore or metal mines, the coal reflecting very little light as compared to the interior of a silver, lead, or gold mine. This resulting contrast was such that when a coal miner looked toward another light source, the glare from a bright reflector with a very dark background tended to blind him. In metal mines, this difference was not as great as the reflected light tended to modify the eyes' pupillary response to bright light. In coal mines, the reflectors were blackened to eliminate this glare. With the flame of a digging mate's lamp providing plenty of sooty carbon, this was relatively quick and easy.

With this ability to carbon a reflector, it was also easy to leave messages on smooth walls for other miners to read and peruse. Underground graffiti about fellow miners, shift bosses, mine owners, and union politics probably flourished only to be mined away by the next shift. In one case, however, it left a more poignant message. (see photo, page 50).

By the time of the Great Depression, electric lights were beginning to challenge carbide lamps for supremacy in the mines. Improvements in battery technology coupled with durable designs from extensive testing both above and below ground had produced an electric light that by 1934 was superior to the carbide cap and hand lamp. As in the case of the carbide lamp just two decades before, comparison testing with the improved electric lamp challenged the miners to open their minds to the new advances in electric lighting.

(Above) An amply protected rescuer preparing to enter the collapsed mine shaft of the Argonaut Mine. Note various light sources and the parakeet.

—Courtesy National Archives

Mine lighting by acetylene cap and hand lamps was becoming outmoded.

In a preliminary study by the Safety Division of the United States Bureau of Mines, it was found that electric cap lamps had nearly three times the candlepower of acetylene cap lamps. In concluding the report, it was stated that

> The carbide lamps have, in the past, given more lumens than most of the electric cap lamps, although many electric lamps have given higher intensities in the center of the beam than the average carbide cap lamp.
>
> The newer types of electric cap lamps apparently provide as much light as carbide lamps, and over a large part of the arena illuminated the foot-candles are greater than those provided by carbide cap lamps. The light uniformity of electric cap lamps is a safety factor helping to reduce mine accidents. . . . In one mine a change from carbide to electric cap lamps very quickly resulted in a decrease of 33.6% in accident frequency.[16]

With the basic technology solved, electric cap lamps became commonplace in mines. Mass production further served to popularize the units as well as lower their purchase price. By the start of the Second World War, only small pit operators and prospectors continued to rely on carbide lamps as their primary source of light underground.

5

Carbide Lamps and Caves

By the mid 1930s, the use of carbide lamps in the mines was becoming a thing of the past. Advancements in the development of the electric light now provided safe, superior illumination at a reasonable cost. As a result, acetylene cap lamps were replaced, and in some cases, banned in favor of the new electric alternative.

Amid the economic hardships of the Depression, this shift in technology forced many carbide lamp manufacturers to scramble for survival. Those that could, merged or diversified their products. Others, with an uncertain market and bleak sales forecast, either sold out or ceased production.

In the years prior to World War Two, the surviving manufacturers struggled to save their shrinking markets. The railroad industry was courted on the benefits of acetylene railroad lamps and lanterns. Hunters, fishermen, and outdoorsmen, albeit a fraction of the total sales as compared to the mining industry in former days, continued their patronage. And somewhere, down in the shuffle of this declining market, carbide lamps found another home.

As the country emerged from the misery of the Depression, Americans found more spare time for their hobbies and pleasure activities. With an available automobile, many took to the out-of-doors, escaping from the city for a week-end in the mountains. Among the various activities of fishing, hiking, camping, and hunting, it was inevitable that some would get curious and enter some of the numerous caves they encountered.

Armed with candles, flashlights, and an occasional carbide lamp, these early explorers learned by doing. Some organized themselves into "speleological clubs," predecessors of the "grottoes" that now form the National Speleological Society. As their experience grew with their interest in caves, these "spelunkers" as they were called, tried and evaluated new equipment and techniques. Fledgling club newsletters were started, and through them, ideas were exchanged concerning exploring, mapping, safety, and equipment. Early on, these cavers recognized the benefits of the carbide lamp and recommended its use in cave exploration.[1] Its popularity grew with the enthusiasm of caving, and many a caver came to treasure his or her brass lamp as the symbol of the good times enjoyed underground.

Acetylene lighting in caves did not originate with these early cavers. As early as 1903 cave chronicler Rev. Horace C. Hovey was touring the hidden corridors of Kentucky's Colossal Cavern with handheld acetylene lights. Indications are that this was a modified "20th Century" brand acetylene bicycle lamp.[2] A few years later, acetylene was being considered at nearby Mammoth Cave.[3] By 1908 there were recommendations to use an acetylene generator outside

(Below) Leo Scott of "Schoolhouse Cave" fame wearing a belt generator by Justrite. (Above left) Four "acetylene" cavers preparing to "go under." (Above right) Burton Faust and Mrs. Howard Watkins preparing to enter Clark's Cave, Virginia, circa 1948. Note lamps and bottoms on the ground. —NSS Photo Archives

the cave and pipe the gas to burners positioned alongside the underground passageways.[4]

In practice, the La Jolla Sea Caves near San Diego, California, were being illuminated with acetylene as early as June 1909.[5] Wonder Cave near Monteagle, Tennessee, also was being shown by acetylene flame. Here, in 1909, a large 100 burner system was installed, later to be increased to 200 burners. The acetylene was generated by the "Aurora" brand generator; the burners were the "Alco" brand manufactured by the American Lava Company of Chattanooga, Tennessee,[6] available for a number of years. Mount Tabor Cave, Ohio, was rigged with a similar system of lighting. Using over 4000 feet of "pipe mains," Mr. William H. Reames lighted his show cave with acetylene from a "Reliance" generator manufactured in Dayton, Ohio.[7] Among the famous Virginia show caves, pretty Massanutten Caverns was illuminated by acetylene flame as early as 1911.[8]

In spite of this early experience with acetylene cave lighting, little sport caving was done in the United States in these first years and hence, acetylene lamps never saw wide use in caves. Indeed, the widely publicized expedition to Carlsbad Caverns sponsored by the National Geographic Society in the mid-twenties, did not use acetylene lamps, but instead employed gasoline lanterns.[9] The famed, yet ill-fated Floyd Collins also used a kerosene lantern on his many solo explorations in cave-rich Kentucky; the rusted remains of one of his supply dumps are preserved deep in Floyd Collins' Crystal Cave to this day. Even during the Sand Cave rescue tragedy in February 1925, carbide lamps were seldom seen among the desperate rescuers at the site.

Several years after the Floyd Collins tragedy, Russell Trall Neville led an expedition into neighboring

Salts Cave. Carrying flashlights, gas lanterns, and candles for illumination, the party explored and slept in the cave, emerging after a then phenomenal 50 hours underground. Only one participant, intrepid George Jackson of Wyandotte Cave notoriety, carried a carbide miner's lamp into the cave. While the others were replacing batteries and broken lantern mantles, George kept his faithful carbide lamp shining cheerfully with a minimum of effort.[10]

In the years following World War Two, sport caving climbed in popularity in the United States. More and more, the carbide cap lamp became the light of choice for experienced cavers nationwide. With their use in pits, crawlways, and rooms, the lamps established a reputation of considerable reliability. Experienced cavers could repair or refill a lamp in total darkness. Just as the miner had found years before, a caver could depend on the warm, soft flame as a welcome source of heat, especially when one was cold, damp, and hours from the entrance.

Some individuals insisted though, that their lamps assumed a mind of their own, acquiring an almost schizoid personality. After hours of reliable service, a lamp might suddenly develop "lampitis," mysteriously failing to burn evenly. Hissing wildly, blowing bubbles through the burner tip, exploding flame around the gasket, or spurting water through the tank door were just some of the bizarre symptoms of "lampitis." Frustration rarely brought a remedy.

Many cavers can recall a trip where the lamp "crapped out" and had to be nursed and cursed back the entire length of the cave only to work wonderfully once outside. Cavers' campfire tales are full of lamp stories. For example, one individual in a fit of utter disgust, after fiddling with his lamp for several hours and stumbling about in near darkness, ripped the thing from his helmet and flung it down a steep canyon—only to have the lamp crash against the distant floor and mockingly burn with a long, smooth, brilliant flame. Another explorer, out of water and in a very dry section of cave, discreetly urinated into his lamp to maintain its flame. It burned gloriously for the remainder of the trip, but never quite functioned as well after that. A fickle mistress indeed.

Then there was the nameless soul outside the entrance of Snedegar's Saltpeter Cave, West Virginia, many years ago, who found he had not cleaned the "spent carbide" out of his lamp from the previous trip. While the others suited up for the long push trip into virgin cave, this hapless fellow proceeded to beat the bottom out of his lamp in the process of removing the hardened, caked residue. No other lamps were available for him to use and he had to sit outside the entrance while his buddies scooped several nice discoveries inside.

(Above) Descending a cave in West Viriginia, circa 1940s-1950s. (Below) A belt generator equipped caver down-climbing flowstone formations in Whiting's Neck Cave, West Virginia, in 1946. —NSS Photo Archives

(Left) *NSS caver with lamp admiring the supposedly authentic signature of George Washington in Madison Saltpeter Cave, Virginia.* —NSS Photo Archives

In other instances, some impatient cavers found carbide lamps to be great as "cattle prods" when bringing up the rear through some long, slow crawlway. Simply inch forward a little too close to the person crawling ahead and wait for the lamp to transmit its message! Unfortunate side effects, however, might mean a boot in the face!

Cavers also found carbide lamps excellent for surveying. By holding the lamp "on station," a surveyor could read the inclination/declination and get a reasonable fix from the compass. The carbon from the flame provided a temporary marker for the next survey station. With such valuable service rendered, it is only fitting that there should be some caves and underground features named after various lamps.

(Below left) Early NSS cavers wearing cloth hats instead of recommended hard hats, circa 1945. (Below) Early NSS cavers tying etriers deep underground. —Both photos courtesy NSS Photo Archives

(Above) William J. Stephenson founder of the National Speleological Society, taping a survey underground about 1950. (Below right) S. A. "Ackie" Loyd attempting cave photography wth flash and carbide flame.
—Both photos courtesy NSS Photo Archive

Take for instance, Justrite Cave, West Virginia, Guy's Dropper Cave in the Tennessee, Alabama, Georgia area, and Auto-Lite Drop in Walt Allen Cave, West Virginia.

As cavers adopted carbide lamps for their explorations, they also modified and improved them. Better tip reamers were devised.[11] Various lamp mounting systems were tried and modified. Someone tried a hard hat balance to counter the front end weight of the lamp.[12] Another devised an ingenious attachment for converting the lamp to a carbide cooking stove.[13]

Others raised valid questions concerning safety. Might not "spent carbide," calcium hydroxide, a strong base, be harmful to cave life? There was no question that dumping it in the cave was detrimental to a cave's beauty; soon, a concensus of responsible cavers agreed to pack all carbide waste out of the caves. The immediate benefits were to the preserved beauty of the cave; only later was it apparent that dumping the stuff into expedition privies significantly improved the neighboring aromas.

Reports of acetylene explosions in cave packs or from damp packets of "spent carbide" were circulated by cave newsletters.[14] Painful burns and in one case, temporary blindness resulted from such explosions. Just like the miners of years before, cavers now learned that it paid to carefully handle carbide waste and get it out of the wet cave environment. Of course, there was always the novice who managed to mistake his carbide for the peanuts and raisins he had packed as a snack. Besides having a strange case of "lampitis," this poor soul could end up with some nasty tasting snacks and a mouth full of burns. (Yes, it really happened.)

By the late 1950's, there were only three carbide mine lamp manufacturers in the United States: Justrite Manufacturing Company, Universal Lamp Company, and Wolf Safety Lamp of America. Though caving was a growing outdoor activity in the country, Universal and Wolf dropped carbide lamps from production, the last "Auto-Lite" dating from April, 1960. Justrite continued as the only domestic producer, electing to drop brass carbide lamps in 1972 in favor of the new plastic lamp. Cavers protested for a return to the old reliable brass lamp but to no avail.

Imported lamps, chiefly the "Premier" from England and the less popular "Butterfly" (Safesport) from the Orient now offer the American caver a steady supply of caving lamps. Still, purists insist that the older American brands are far superior. Old "Guy's Droppers," "Auto-Lites," "Justrites," and possibly an occasional "Wolf" are still available for cavers, but they command premium prices in many areas. Some individuals, referred to in some quarters as "lamp tramps," now collect these older lamps for trade as well as fun. Not necessarily cavers, these folks insure that carbide lamps will be preserved and studied for posterity and kept as a memento to an era that once was.

6

Manufacturers and Their Lamps

It has now been more than a decade since the last brass carbide lamp was made in the United States. The demand for these lamps, admittedly small when compared to past years, is now met by imports from England and the Orient. Good, reliable domestic brands are still available . . . at flea markets and antique shops and occasionally, a new, unfired "Justrite" or "Auto-Lite" will "turn up" in some country store.

Yet, for many years, there were a prolific number of American lamp manufacturers, all seeking to carve their niche in the mine lighting market. Their ideas and inventions, their time and money, and their successes and failures have provided collectors and students of the subject with many interesting insights into the acetylene lamp industry. However, many questions remain.

What follows is a collection of individual company histories. Counting only a few exceptions, none of these companies is in existence today. The information and research gathered, often distanced by the span of half a century or more, has been difficult to trace. Yet, it is hoped that by itemizing each company, its agents, products, and other relevant facts, a rough picture of the industry as a whole—how it evolved, grew, and finally succumbed—might be portrayed. Additionally for the student and collector, such information should be helpful in further advancement of the study and collection of miners' carbide lamps.

A note of caution is warranted in regard to dates. It is very difficult to date most lamps exactly. If not adequately documented, all precise dates should be suspect. In such cases, it is more accurate to describe a particular lamp as "circa 1918-1919" rather than try to pinpoint a specific date based on hunches or loose sales data. Catalogs and trade advertising can also be deceptive regarding dates and designs as many lamps made during the peak World War One years were sold in the 1920s long after production had stopped.

The manufacturers are listed alphabetically and include relevant documentation after each section. A closing section describing many of the "loose ends" of carbide lamp history—unknown lamps, companies with no known product, advertising brochures, hunches, etc.—is included for future researchers and "lamp tramps." Although the zenith for these curious mining lamps has passed, many puzzles remain for those interested in these relics of America's industrial and recreational past.

The line drawings of lamps in this and subsequent chapters were graciously provided by Dr. Wendell Wilson of Tucson, Arizona.

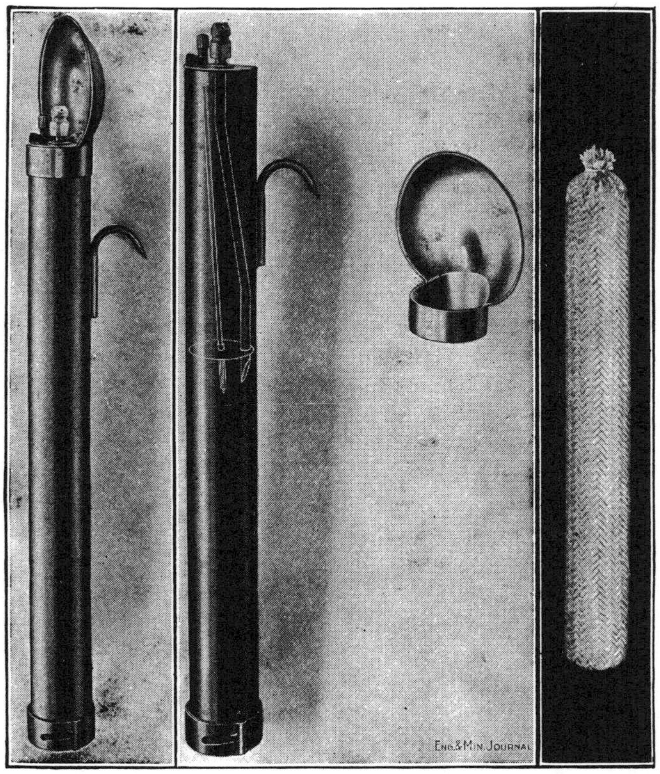

The Arnold Carbide Candle

WILLIAM AINSWORTH & SONS
Denver, Colorado

Of all the acetylene mining lights developed in the United States, none is more curious than the "Arnold Carbide Candle," manufactured by William Ainsworth & Sons. Though there were several "carbide candle" patents, only Ralph R. Arnold's 1912 patent bore fruit in the form of an actual mine lighting device.[1]

Arnold, a mining engineer from Cripple Creek, Colorado, first tried his novel prototypes in the ore mines of Colorado about 1911. Ainsworth was contracted to manufacture the "candles" soon after Arnold received his patent. The "acetylene candle" consisted of a

> tubular body of 22-gauge brass, 1¼ in. in diameter by 12 in. long, with a diaphragm in the center dividing it into a water (upper) chamber closed by a revolving shutter at the top, and a carbide (lower) chamber closed by a gasket-lined cap at the bottom. A water-control valve is placed in the diaphragm and a gas tube extends from the carbide chamber through the diaphragm at the top and terminates in a burner.
>
> The cartridge, containing two ounces of calcium carbide, sufficient to produce a 4-cp. light for four or more hours, is carried in the carbide chamber, the fabric being designed to filter the gas, admit of the expansion of the carbide and contain the consumed carbide. . . .[2]

With the growing popularity of the acetylene cap lamp, this peculiar, probably awkward device soon became just another curiosity and production ceased.

The Ainsworth Company as it is now known, was founded by William Ainsworth, a watchmaker born in England and raised in Wisconsin. In 1874 he arrived in Denver and soon established a small jewelry shop in Central City. It was not long before Ainsworth acquired a reputation for his excellent repair work on assay balances, so essential to the mining industry.

In 1880 he moved back to Denver and opened shop at Sixteenth and Larimer streets. There, he began manufacturing his own balance known as the "Button Balance," the first of many, quality analytical balances. Business increased and in 1894 the company began producing the famous Brunton Pocket Transit.[3]

With William Ainsworth's death in 1917, his sons, Robert G. and Alfred W. assumed control of the company. Incorporated in 1923, The Ainsworth Company continues to produce quality, precision products today.

1. See Letters Patent No. 1,043,039 of October 29, 1912 to Ralph R. Arnold.
2. "Arnold Carbide Candle", *The Engineering & Mining Journal*, January 18, 1913. p. 185.
3. Ellsworth C. Mitick, "A History of Mining Machinery Manufacturers," MS thesis published in *Colorado Magazine*. Vol 25. No. 2, March 1948. pp. 80-83.

Also see: A. Cressy Morrison, "Acetylene Mining Lighting," *Mining Science*, April 1915. p. 31.

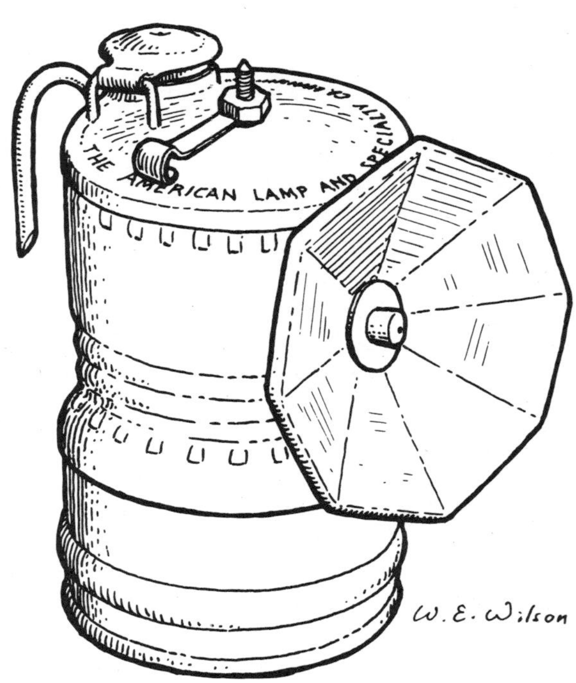

An "American" Lamp
After Pohs

AMERICAN LAMP & SPECIALTY COMPANY
Evansville, Indiana

This short-lived company manufactured the "American" carbide cap lamp. James H. Powers, company president, patented a water-feed valve for carbide lamps on July 9, 1918.[1]

It is not known with certainty when the first of these "American" lamps were made but by 1922 Powers had the company established and doing business at 1204 Main Street. By 1924 Powers had departed and Edward J. Miller, company treasurer under Powers, became president.[2] As this is the last listing for the company in the city directories of Evansville, it is probable that they soon ceased production and closed shop.

It has not been established if Edward J. Miller was ever connected with the Edward Miller & Company of Meridian, Connecticut, manufacturers of early miners' lamps.[3] In business before 1909, this firm was not engaged in carbide lamp manufacture, but instead, made oil-wick lamps.

1. See Letters Patent No. 1,272,315 of July 9, 1918, to James H. Powers.
2. City Directory of Evansville, Indiana, 1922, 1923, 1924.
3. Thomas Register of Manufacturers, 1909.

Anton cubical-top

GEORGE T. ANTON & BROTHER
Monongahela, Pennsylvania

The Antons, prolific oil-wick cap lamp producers, attempted to enter the carbide cap lamp market around 1915. John B. Anton patented several ideas on his "square" Anton cap lamp just weeks before George Shanklin of the Shanklin Manufacturing Company patented his very similar design on the "Squarelite" Guy's Dropper.[1] Another Anton lamp, this one resembling the "Maple City" cap lamp, has been found indicating some variety of style on the Anton's part.[2]

The Antons started their oil-wick lamp business around 1898 and were located at 7th and Railroad streets.[3] They made a number of oilwick lamps including the "Star," the "Globe," and the "U.S. Eagle" brands. The company appears to have ceased lamp production about 1923.[4]

1. See Letters Patent No. 1,187,481 of June 20, 1916, to John B. Anton. Though often referred to as the "Anton" carbide cap lamp, the Antons labled the light like many of their oil-wick lamps, the "Star."
2. *Underground Lamp Post*, Spring 1981. p. 2
3. City Directory of Monongahela, Pennsylvania, 1909, 1916.
4. Thomas Register of Manufacturers, 1923. Trade advertising at this time shows the Antons dropping out of many periodicals. The 1924 Thomas fails to show the Antons as being in the lamp business.

See also U.S. Trademark of August 25, 1908 to George Anton which documents the use of the name "Star" from September 5, 1898.

Anton

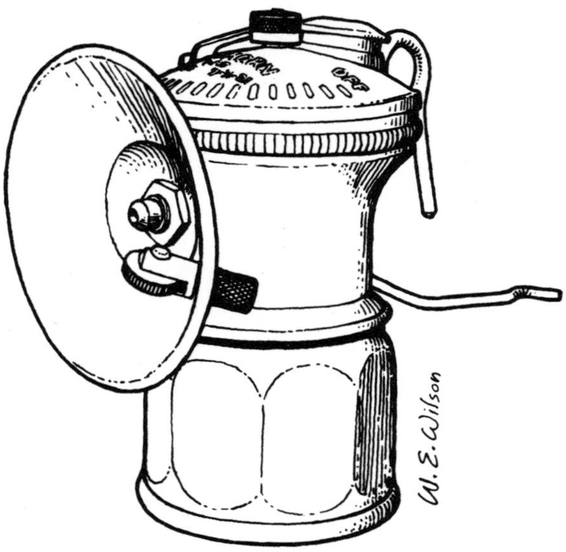

The Elkhorn

The Buddy

ASHMEAD MANUFACTURING COMPANY
Ashland, Kentucky

The Ashmead Manufacturing Company produced two brands of miners' carbide cap lamps, the "Elkhorn" and the "Buddy." Marketed in both brass and nickel plate, the "Elkhorn" was Ashmead's first lamp, coming in either cap or superintendent's models. Some time later, in order to generate a wider appeal, the "Buddy" logo was adopted for the product.

The "Elkhorn" came packaged in a blue and gray box; a red and yellow box was used for the "Buddy." Both lamps, though never produced simultaneously, were packaged six to a carton, twelve cartons to a case, and shipped to such countries as New Zealand, Australia, Bolivia, Mexico, and Peru. Interestingly enough, each lamp was estimated to contain ten cents worth of materials and to cost an equal amount to assembly. The lamps were billed to the jobber at sixty cents each and sold at a retail price of one dollar. Nickel plate cost a quarter extra.[1]

The Ashmead Manufacturing Company was started sometime during World War One. Ben Williamson, president, bought the Meyer and Philip Stein patent of May 3, 1921, and incorporated its water-feed innovations into the "Elkhorn."[2] Later, Ben Williamson's patent No. 1,716,809 added improvements to the "Buddy."[3]

The Ashmead Manufacturing Company was located at 125 16th Street. In 1925 Davis Geiger became president of the company which was owned by the Williamson family. The company closed operations in 1933.[4]

1. Much of the information on the Ashmead Manufacturing Company was obtained from personal interviews with Mr. Davis Geiger, a former employee and president of the company. Specifically, see the 1976 interview with James Lackey, published in the Fall, 1977 *Underground Lamp Post*, and this author's correspondence with Mr. Geiger at Del Ray Beach, Florida, December 1980.
2. See Letters Patent No. 1,376,654 of May 3, 1921, to Meyer Stein and Philip Stein.
3. See Letters Patent No. 1,716,809 of June 11, 1929, to Ben Williamson.
4. *Op. cit.*, Lackey interview, Clemmer interview.

FREDERIC E. BALDWIN
JOHN SIMMONS COMPANY
ZAR MANUFACTURING COMPANY
New York, New York

Of all the designers and inventors working on acetylene mine lamps, Frederic E. Baldwin was probably pre-eminent. With his 1900 patent for improvements in an "acetylene-gas lamp,"[1] Baldwin began a career that would soon make his "Baldwin Pit Lamp" the most popular carbide lamp in the United States. From his shop at 105 Centre Street in New

The Baldwin Lamps

York, he founded a company which would be the seed to an industry that would eventually manufacture millions of lamps and span more than 60 years.[2]

The first lamps, other than the early patent models, appear to have been fabricated and marketed by Albert H. Funke in the spring of 1900, several months before Baldwin was granted his first lamp patent.[3] Working under license with Baldwin, Funke marketed a cumbersome, nickel-plated lamp called the "Full Moon."[4] This relationship between inventor and manufacturer did not last for many months because Baldwin lamps were soon being sold by the Ingersoll-Sergeant Drill Corporation. This arrangement apparently lasted several years as evidenced by trade advertising, while Baldwin manufactured his lamps and struggled to meet a growing public demand and curiosity for his products.

Probably in need of working capital and desiring better marketing and manufacturing arrangements, Baldwin contracted in 1906 with the John Simmons Company, a large plumbing and machinery distributor, to manufacture and market his lamps. This can be considered "a red letter date" in carbide lamp history, for it was at this time that the merits of the acetylene lamp for mine application began to be accepted in the United States.

The first Baldwin-Simmons lamps, although the best product for its time, could be considered crude and primitive by later standards. A poorly controlled wire water-feed mechanism was the greatest design challenge for the proper functioning of the lamps. This problem would continue to concern Baldwin in future patents.

Over the next eight years, the "Baldwin Pit Lamp" or "Simmons Lamp" enjoyed enormous popularity in both metal ore mines and coal mines. By 1913 the company was boasting that "80% of the Carbide Lamps in use are BALDWIN lamps."[5]

Success spurred Baldwin and Simmons to market a growing variety of cap lamps for miners. The company had revised the early pinch-waist cap lamp of 1906 into the modern "No. 32" model. This popular lamp along with other hand and full shift lamps continued to dominate the market until the coming of the First World War in Europe.

One of the stranger Baldwin lamps was the "No. 33" which possessed a bent reflector that both focused light straight ahead as well as at a spot on the floor immediately in front of the miner. In 1914 the company offered a "surveyor's lamp."[6] This novel lamp utilized a wide, curved reflector that acted as a bright backdrop in order to better visualize a suspended plumb bob.

Throughout this prosperous period, Baldwin continued his experimentation and design work on carbide lamps. Many of his patents were concerned with

1906 *1914*

THE BALDWIN

The Miner's Lighting Bug

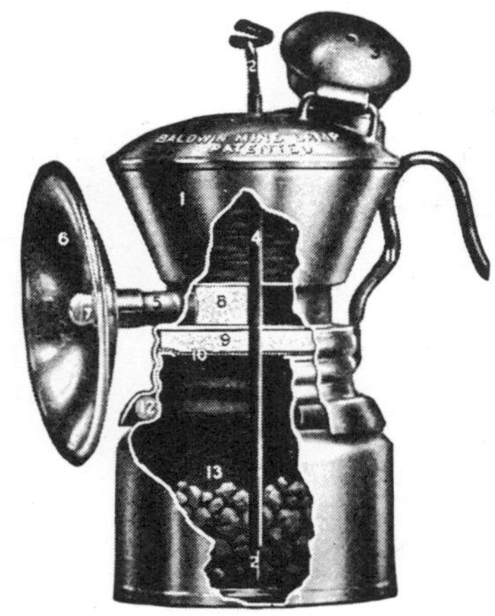

Still Satisfactory After Eight Years of Service Without Any Radical Changes

Does this mean anything to you? If you have kept in touch with the various lighting devices that have been put on the market in this space of time, it proves that what we have claimed for the Baldwin is true, it operates under the only known method for accurately controlling the generation of gas in a carbide pit lamp.

> The valve lamp has been offered in a dozen different forms in the past five years. Why? Simply because a valve feed lamp is not practical and it has been found necessary to make continual changes to overcome difficulties that have cropped up when the lamps have been put in use.

Compare the Baldwin No. 32 lamp of to-day with that of 1906. There has been a graduai, steady development of the Baldwin Lamp; those two illustrations do not tell the history of the Baldwin—simply show the advance in appearance. The advance in strength—serviceability and efficiency cannot be shown in a picture but it's well told in the lamp. Many of these changes were made at the suggestions of practical men who used the lamps daily in their work. We are always on the lookout for improvements that will add to the efficiency of the Baldwin Lamp, but unless a change will stand up under the rigid tests we put it to, it is not put in a stock lamp. We do not sell experimental lamps. That is why the Baldwin Automatic Water Feed has stood unequaled for over eight years.

> In the Baldwin, the water flow is determined by the gas pressure itself, which checks or holds the water in reserve until it is required thus insuring a steady, even flame at all times, whether the miner is standing still or swinging a pick. The only attention a Baldwin Automatic Feed lamp needs is a turn of the raking wire a few times a day.

More lamps of this feed are in use than all others combined and further proof of the value of the Baldwin Automatic Feed is found in the number of valve feed lamps from which the valve has been removed and a Baldwin water tube and raking wire inserted.

> These are facts, but we invite you to investigate the Baldwin Lamp—find out what it can do—has done—and will do.

Our Proof Is the Lamp Itself

Equipped With or Without Sparker Lighters

Lamps are guaranteed satisfactory to user—get one at your dealer's today. If you don't find it just as represented, take it back and your money will be refunded.

John Simmons Company

100 Centre Street **New York**

BRANCHES: 268A Market St., San Francisco, Cal. 140 Bleury St., Montreal, Can.

perfecting the water-feed mechanism and designing a better reflector-mounted flint striker.

Sometime in 1913 Frederic E. Baldwin, for reasons unknown, left the John Simmons Company and formed the Zar Manufacturing Company, trading at 320 Broadway, New York City. This "Zar" cap lamp that Baldwin now began to manufacture had just begun production at Simmons and evidently was given to Baldwin as part of his share of the business. Armed with several distinctive patents and a new name, Baldwin believed that his newest lamp was the "best lamp built."[7] Baldwin probably envisioned the "Zar" as a strong competitor with the established "Baldwin Pit Lamp" manufactured by Simmons, but it was never to be. By 1918 the Zar Manufacturing Company was defunct and Frederic Baldwin had formally sold 17 of his patents to Simmons.[8] Business directories did, however, continue to list the "Frederic Baldwin Manufacturing Company" at 320 Broadway until at least 1924.[9]

If any individual can be called the "Father of the Carbide Miners' Lamp," it is Frederic E. Baldwin. As an inventor and designer of acetylene lighting, he had few peers. Without a doubt, his pioneering work in mine lighting contributed enormously to safer work conditions underground and increased coal and ore production to the benefit of both miner and owner.

The John Simmons Company succeeded in marketing the Baldwin lamp where its inventor had failed. After the agreement with Baldwin in 1906, the company became the dominant acetylene mine lamp manufacturer in the United States. In addition to its own brand of miners' lamps, Simmons appears to have "private labeled" a number of lamps for other mining supply companies. Under President Charles H. Simmons, the company marketed lamps throughout the United States as well as in Cuba, South America, Latin America, Australia, and New Zealand. Simmons' trademark, "The Miner's Lighting Bug" became known worldwide.[10]

When the John Simmons Company bought the 17 Baldwin patents on October 11, 1918, the company itself was in the process of closing carbide lamp operations.[11] Wilbur A. Cochrane, an inventor and employee with Simmons as far back as 1910,[12] had formed the Dewar Manufacturing Company with Francis H. Coffin, formerly associated with the Scranton Acetylene Lamp Company, and John M. Brock, another lamp designer and inventor.[13] The year of 1919 saw Dewar acquire one of the original Baldwin patents owned by Simmons.[14] By 1921, with Simmons now almost completely out of the lamp market, Dewar acquired another of the critical Baldwin patents.[15]

John Simmons Company also sold one of the Baldwin patents to Jacob Sherman of the Universal Lamp Company.[16] Thus, Frederic E. Baldwin, via John Simmons Company, provided part of the technical foundation to the manufacturer of the popular "Auto-Lite" and "Guy's Dropper" as well as to the manufacturer of the "Sun Ray," "I-T-P," and "Dew-R-Lite."

1. See Letters Patent No. 656,874 of August 28, 1900, to Frederic E. Baldwin.
2. See accompanying chart.
3. *Scientific American*, June 23, 1900.
4. *Engineering and Mining Journal*, September 15, 1900. pp. 312-13.
5. *Ibid.*, January 4, 1913. p. 9.
6. *Ibid.*, February 21, 1914. p. 425.
7. See enclosed Zar advertising, circa 1914.
8. *Digest of Assignments*, United States Patent and Trademark Office. October 11, 1918—Volume B.

   For those interested, these patents are listed numerically below:

821,580	1,077,582	1,115,157	1,237,449
994,365	1,083,427	1,174,035	1,237,450
1,066,241	1,094,358	1,174,036	
1,070,000	1,099,908	1,174,037	
1,075,960	1,101,982	1,174,112	

9. New York Business Directory, 1918-1924.
10. See U.S. Trademark No. 105,876 of August 24, 1915, to John Simmons Company, Charles H. Simmons, President.
11. This is apparent by the marked drop in Simmons' trade advertising after 1916.
12. *Acetylene Journal*, July, 1910. p. 30.
13. See Dewar Manufacturing Company elsewhere in this volume. The "I-T-P" trademark was granted Dewar on July 18, 1916, and had been in use with the company since October, 1915.
14. *Digest of Assignments*, United States Patent and Trademark Office. December 11, 1919—Volume S. This was patent No. 1,094,358.
15. *Ibid.*, May 25, 1921. Patent No. 1,083,427.
16. *Ibid.*, December 29, 1920. Patent No. 821,580 and reissue No. 13,542.

See also: *Mining and Engineering World*, July 11, 1914, p. 65; and *Engineering and Mining Journal*, October 12, 1901, p. 465-66.

PIONEERING INTERRELATIONSHIPS OF FREDERIC E. BALDWIN WITH OTHER LAMP MANUFACTURERS

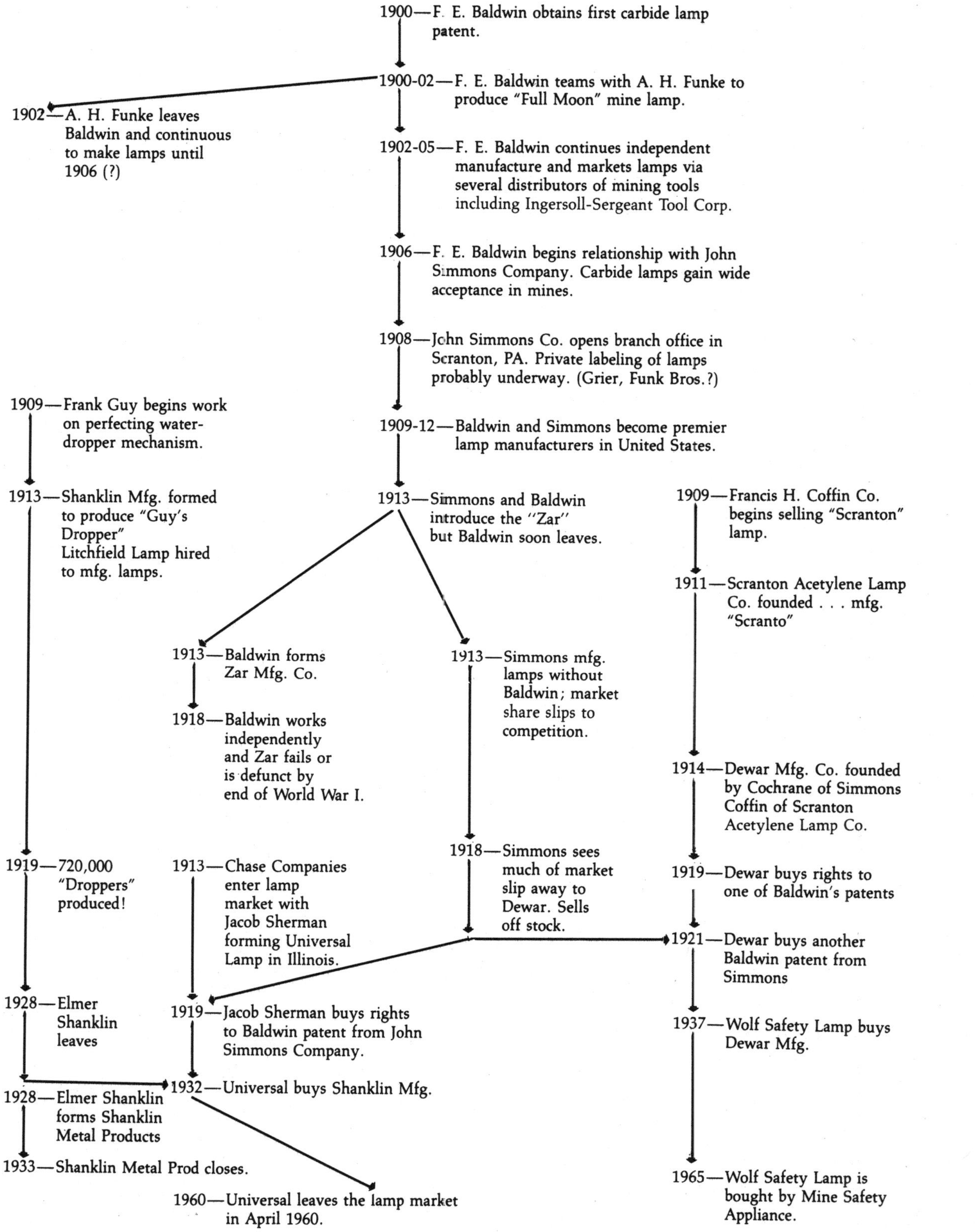

BEALL BROTHERS SUPPLY COMPANY
Alton, Illinois

Recently located trade advertising has raised speculation that Beall Brothers may have manufactured and marketed the "Red Star" acetylene mine lamp.[1] The company is widely known to have made the oil-wick lamps for a number of years and indeed, may have entered the acetylene lamp market, if only for a while. Although there are no known lamps to confirm this,[2] at least the advertising supports this conjecture.

The lamp's similarity to other lamps, notably the "Maple City" brand, leads some collectors to speculate that the "Red Star" could have been manufactured by the Monmouth, Illinois, company and marketed under the Beall name. Admittedly this is tenuous and there is no hard evidence for its support.

Beall Brothers dates back to at least 1910 as a manufacturer and marketer of "miners' tools and supplies."[3] In 1923 Carl W. and John Beall were the listed principals in the business which was located at 123 Illinois Avenue; the shop was at 129 Indiana Avenue. By 1929, J. W. Hubbard was president and J. W. Beall was vice-president.[4]

1. Larry Peterson, personal communication, February 26, 1980. Keystone Catalog, 1921. p. 1293.
2. *Ibid.*, September 6, 1981.
3. City Directory—Alton, Illinois, 1910.
4. *Ibid.*, 1923-29.

FRED R. BELT COMPANY
Chicago, Illinois

The "Lu-mi-num" brand carbide lamp manufactured by the Fred R. Belt Company represented a rather sophisticated deviation from the standard acetylene mine lamp design. Fabricated from an alloy of aluminum and copper, the lamp body was die-cast instead of being built by the usual die-stamp and solder techniques employed by the other lamp companies.

There are a few screw thread "Lu-mi-num" lamps in existence; however, the majority utilize the double cam lock to secure the lamp bottom. It is probable that the screw bottom lamps were developed first as a trial pattern before the distinctive "Lu-mi-num" lamp was patented in 1926.[1]

The Fred R. Belt Company secured the "Lu-mi-num" trademark on May 20, 1924, stating that they had used the "Lu-mi-num" name since September 19, 1923.[2] The unusual ribbed, heat dissipating design was patented February 5, 1924, and the lamp and improvements were patented two years later.[3] Both hand and cap lamps were manufactured under the "Lu-mi-num" label.

Located at 355-361 West Ontario Street, the Fred R. Belt Company also designed and built "special-use" machinery. The company left the mine lamp market about 1933.[4]

1. See Letters Patent No. 1,612,694 and 1,612,695 of December 28, 1926, both to Fred R. Belt.
2. See U.S. Trademark No. 184,272 of May 20, 1924 to Fred R. Belt.
3. See Design Patent No. 63,868 of February 5, 1924 to Fred R. Belt.
4. Thomas Register of Manufacturers, 1933, 1934.

See also: *Engineering and Mining Journal Press*, October 10, 1925, p. 593.

BOESCH LAMP COMPANY
San Francisco, California

Although miners' "acetylene gas lanterns" were advertised by Boesch Lamp Company, it is not certain how many of the lamps were sold and indeed, if any survive to this day. The only information is in the advertising.[1]

Constructed of brass with a nickel plated or steel finish, the "Boesch" lantern stood 7½ inches high and was equipped with a bail, hand loops, and belt hook. It was advertised as the "only Lamp that meets the requirements of Miners and Mine Superintendents."

The company was established in 1869 and manufactured a wide variety of railroad, automobile, marine, and home lamps. In 1915 Boesch is listed as a "manufacturer of acetylene lamps" doing business at 1135 Mission Street. From as far back as 1895, the business was located on Mission Street until 1926 when listings in the San Francisco City Directory fail to show the Boesch Lamp Company. It is not known what relationship the "1569 Grove St." address had with the company since this address was not listed with the company.[2]

On a related point, in 1912 the Hiller Lamp Company briefly opened at 143 2nd Street with Stanley I. Hiller as proprietor. Whether he was related to the president of Boesch, J. R. Hiller and the secretary-treasurer, K. A. Hiller, is a matter of conjecture. It is not known if the Hiller Lamp Company manufactured acetylene lamps, and further research will have to be done before any connection might be established.[3]

1. *The Underground Lamp Post*, Vol. III, No. 6, Spring, 1981. p. 6.
2. Thomas Register of Manufacturers, 1915.
   City Directory of San Francisco—1895-1926.
3. *Ibid.*, Directory of 1912.

Further research has turned-up a Henry Boesch Company at 275 Greenwich, New York City, doing business as a lamp manufacturer circa 1907. (Business Directory of New York City, 1907.)

BRILLIANT SEARCH LIGHT MANUFACTURING COMPANY
Chicago, Illinois

One of the earliest portably acetylene lamps to emerge in the United States was the "Brilliant Search Light," manufactured by Rudolph C. Kruschke of Duluth, Minnesota. Kruschke was a dealer in guns and hunters' supplies in Duluth and patented his lamp in 1902.[1] This lamp was one of the first waist-belt generator, head-burner-reflector lamps.

Dating back to 1899,[2] the lamp was primarily advertised to the outdoorsman for use in such activities as fishing, camping, canoeing, and hunting. Yet, as early as 1903, miners were being advised of the "Search Light's" effective use in their work,[3] although its use in the nation's mines was never very popular when compared with the miners' acetylene cap lamps.

In later years, Kruschke moved the business to Chicago, advertising his lamps in many periodicals including the trendy Sears and Roebuck Catalogue. The lamps continued to be available for many years until the outbreak of World War Two.

1. See Letters Patent No. 711,871 of October 21, 1902, to Rudolph C. Kruschke.
2. *Acetylene Journal*, September, 1903. p. 90.
3. *Ibid.*, p. 90.

See also, Letters Patent No. 1,093,339 of April 14, 1914, to Rudolph C. Kruschke concerning improvements in his "mining headlight" for the "Brilliant Search Light."

T. M. CHANCE
Philadelphia, Pennsylvania

Technological advances in acetylene mine lamps brought a variety of changes, even aberrations to the design of the lamps. In the years prior to World War One, there was a small school of inventors who sought to perfect an acetylene safety lamp, incorporating both the superior illumination of acetylene and the gas detection features of the contemporary safety lamp.

One of these experimenters, T. M. Chance, built such a lamp and ran a series of tests on it.[1] Admittedly not as bright as the popular cap lamp, the Chance lamp did receive at least one positive review from one observer. "There is a new acetylene safety lamp that gives promise of being an admirable working lamp as well as an excellent fire damp and black damp detector."[2]

Although discussed at several meetings of the American Institute of Mining Engineers, the Chance safety lamp never was more than a curiosity and after a while, disappeared from consideration.[3]

1. Burrows, R. P., "Illumination of Mines," *Transactions of the American Institute of Mining Engineers*. Vol. 54. p. 34. 1916.

Chance, Edwin M., "Portable Miners' Lamps," *Transactions of the American Institute of Mining Engineers*. Vol. 55. p. 241.

2. *Ibid.*, Chance. p. 243.
3. See Letters Patent No. 1,323,618 of December 2, 1919, to T. M. Chance.

COOLAMP COMPANY
Benton Harbor, Michigan

Very little is known concerning the Coolamp Company; it is not even known if they manufactured miners' carbide lamps. Most of the evidence on the Coolamp Company centers on the patent of Raymond A. Mull, who patented an acetylene cap lamp on February 2, 1926.[1] He had first filed his application for this patent on March 22, 1922 and on June 12, 1922, he assigned all rights and privileges of the patent, still unissued, to the Coolamp Company.[2]

As none of the lamps are known to exist and inquiry locally at Benton Harbor has revealed nothing concerning the Coolamp Company or Raymond Mull, the existence and circumstances concerning the use of his patent and any surviving lamps remains a mystery.

1. See Letters Patent No. 1,571,554 of February 2, 1926, to Raymond A. Mull.
2. *Digest of Assignments*, Vol. M 48, p. 104.

JOHN S. CUMMINGS & COMPANY
Tunnelton, West Virginia

Very little is known concerning the John S. Cummings Company or the "Standard" acetylene mine lamp they manufactured. Local inquiries around Tunnelton located a few descendents of Mr. Cummings but scant knowledge of the company or its products.[1]

The "Standard" lamp is known only through several early advertisements placed in mining journals in 1902-03. Primitive in design, it contained a centrally located carbide chamber. Water, located beneath this carbide container was fed via a wick fastened in a cork filter, upward against the pull of gravity.[2]

The lamp was three inches tall, one and one half inches in diameter and was designed to burn 2-3 hours on 1¼ ounces of carbide. It weighed 5 ounces and was priced at one dollar.[3]

James W. Paul, Chief Mine Inspector of West Virginia and later author of the miners' circular, "Notes on Miners' Carbide Lamps," examined the "Standard" and stated, "Its general adoption in all mines in which explosive gases are not found would very greatly benefit the health and safety of the underground employees."[4]

There are apparently no patents assigned to Cummings concerning this lamp. Unknown in any collection today, it was one of the earliest American acetylene mine lamps.[5]

1. Larry Peterson, personal communication. September 6, 1981 and February 26, 1980.
2. *The Michigan Miner.*, November, 1902. p. 27.
3. *Mines and Minerals*, September 1902-February 1903. Various advertisements.
4. *Ibid.*, September 1902. p. 62.
5. *Op. cit.*, Peterson, February 26, 1980. This researcher-collector indicates he has information that dates Cummings' "Standard" lamp back to 1899.

See also, *The Engineering and Mining Journal*, October 4, 1902. p. 453.

According to the *Michigan Miner*, November 1902, Mr. A. C. Stalknecht was the general sales agent for Michigan and Wisconsin. Stalknecht's business was located in Saginaw, Michigan, but it is not certain if his business in the "Standard" lamps was a prosperous one.

DEWAR MANUFACTURING COMPANY
Brooklyn, New York

The Dewar Manufacturing Company was formed around 1914[1] by Wilbur A. Cochrane, an inventor formerly with John Simmons Company, Francis H. Coffin, formerly associated with Scranton Acetylene Lamp Company, and John M. Brock, lamp designer and inventor. Located at 28-30 35th Street in Brooklyn, the company began manufacturing the "I-T-P" (It's Trouble Proof) cap and hand lamps.[2]

By 1918 Dewar had expanded its production to include the new "Sun Ray" cap lamp and had established a Canadian office in Toronto, Canada. Negotiations between Dewar and John Simmons Company at this time culminated in the purchase from Simmons of two original Baldwin patents.[3]

Besides purchasing patent rights from others, Brock and Cochrane patented a number of their own ideas for carbide lamps including improvements in flame protectors and a wind break, water-feed valves, and cap lamp attachments.[4] The "Dew-R-Lite" hand lamp, manufactured after the "I-T-P" and "Sun Ray" had been introduced, utilized many of these improvements.

When Brock died about 1927, Cochrane was named executor of his estate and assumed total control of the company.[5] Dewar continued to manufacturer acetylene lamps throughout the Depression until they were bought by the Wolf Safety Lamp Company of America, Inc.

1. The first mention of the Dewar Manufacturing Company comes from the *Digest of Assignments*, U.S. Patent and Trademark Office. It seems the Alonzo Roach patent for a reflector-mounted flint striker, No. 1,068,931 of July 29, 1913, was sold to Percy G. Dick on April 7, 1914, who the same day sold his rights to Dewar. Thus Dewar was at least a legal entity in 1914, through the exact date they began manufacturing lamps is not known.
2. See U.S. Trademark No. 111,454 of July 18, 1916 to Dewar Manufacturing Company, the "I-T-P" name having been used since October 1915.

CARBIDE ITP LAMPS

IT'S TROUBLE-PROOF

FLOAT FEED

Patented

The Simplest of All Lamps

It is less complicated than other lamps. You simply charge it with carbide and water, open the valve full wide, light and it burns and it keeps on burning until the charge is completely used up. No raking. No shaking. No adjusting.

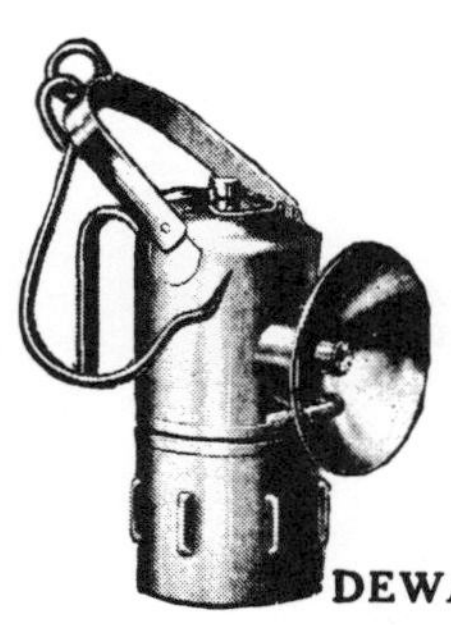

Care of this lamp is equally simple. May we send you evidence of this in the form of a sample lamp? No obligation incurred on your part.

DEWAR MANUFACTURING CO., Inc.

28-30 Thirty-fifth St.
Brooklyn, N. Y.

100 Wellington St., W.
Toronto, Ontario

3. *Digest and Assignements*, United States Patent and Trademark Office. December 11, 1919. This was patent No. 1,094,358. May 25, 1921. This was patent No. 1,083,427.
4. For a complete count of the Brock and Cochrane patents, refer to the Patent Index.
5. *Op. cit., Digest and Assignments*, November 10, 1928.

FUNK BROTHERS MANUFACTURING COMPANY
Chicago Heights, Illinois

Recently, a "Funk Bros." brass cap lamp was found by a collector; it strongly resembles the Baldwin "No. 32" pit lamp.[1] Little is known concerning the Funk lamp, but it appears to be another case of private labeling from another manufacturer, most likely the John Simmons Company of New York.

Research into the Funk Brothers Manufacturing Company shows that the firm was a large dealer in miner's tools and coal drilling equipment and certainly carried carbide lamps in their inventory. As manufacturers, it is not improbable that they may have manufactured lamp prototypes to test the market. One piece of evidence supports this: *The Thomas*

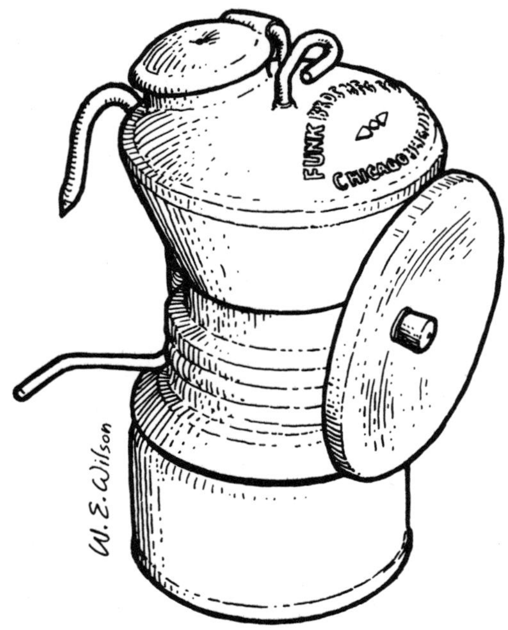

Funk Brothers Lamp

The Full Moon

Register of Manufacturers for 1914 lists Funk Brothers Manufacturing Company under acetylene lamp manufacturers.

1. *The Underground Lamp Post*, Vol. III, No. 6, Spring, 1981. p. 1.

A. H. FUNKE
New York, New York

As an associate of Frederic E. Baldwin, Albert H. Funke was one of the pioneers of carbide mine lamps. As a manufacturer, he was already producing the "Indestructible" oil-wick cap lamp for miners. By 1899 he was exploring acetylene lamp designs and had begun fabricating and marketing his "Crescent" acetylene automobile lamp.[1] In February 1900 he added the "Autolyte" lamp to his product line.[2] By June, he was making and selling the new "Baldwin" bicycle carbide lamp.[3]

After Baldwin received his patent in August of 1900, Funke obtained a license to the Baldwin patent rights to manufacture a mine acetylene lamp.[4] Known as the "Full Moon" mine lamp,[5] this lamp stood 5 inches tall, weighed 9 ounces, and burned for 4 hours. A handle could be attached to a catch on the back of the lamp for carrying underground.

Funke continued operations until at least 1906,[6] concentrating mainly on automobile and bicycle markets after Baldwin began marketing his own mine lamps with the Ingersoll-Sergeant Company in 1901.[7]

In 1900, Funke's shop was located at 101-103 Duane Street.[8] By 1906, he had moved to 83 Chambers Street.[9] Though no positive connection has been established, Funke could have been associated with the Autolyte Manufacturing Company of 26 Warren Street, circa 1909, makers of "automobile, acetylene mining lamps."[10]

1. See U.S. Trademark No. 50,572 of March 20, 1906, to Albert H. Funke. Funke states in this document that he has been using the "Crescent" trademark since 1899.
2. See U.S. Trademark No., 53,270 of May 29, 1906, to Albert H. Funke. Funke states in this document that he has been using the "Autolyte" trademark since February 27, 1900.
3. *Scientific American*, June 23, 1900.
4. *Engineering and Mining Journal*, September 15, 1900. pp. 312-3.
5. *Ibid.*, p. 313.
6. *Thomas Register of Manufacturers*, 1906, 1907. Although Funke is no longer listed in 1907, it seems probable that his exit from the mining lamp market occurred much earlier when Baldwin began marketing with Ingersoll-Sergeant and then later sold patent use rights to John Simmons Company.
7. *Op. cit.*, October 12, 1901.
8. *Op. cit.*, *Scientific American*.
9. *Op. cit.*, Trademark No. 53,270.
10. *Thomas Register of Manufacturers*, 1909. p. 1146.

GEM MANUFACTURING COMPANY
Pittsburgh, Pennsylvania

The successor to the "Grier Bros." lamp was the "Gem," manufactured by the Gem Manufacturing Company. Like the "Grier Bros." the "Gem" was optionally equipped with the J & T tip cleaner. Manufactured in both brass and nickel finishes, the "Gem" was available in either cap or hand lamp models.[1]

Gem Manufacturing Company was incorporated in 1889 and located at 1229-1243 Goebel Street. E. S. Fownes was president of the company.[2] The "Gem" name also was identified with "pump oilers, hand oilers, supply cans and metal stampings" which the company manufactured.[3]

The "Gem" carbide lamp was evidently first marketed in 1925 and sold for a number of years.[4] It is not certain when the "Gem" lamp was halted in production, but a 1950 listing of the company shows J. A. Fownes as president.[5] There is no mention of lamps in this listing. The company continued in existence until 1961.[6]

1. *Keystone Mining Catalog*, 1925. p. 579.
2. *Pittsburgh City Directory* for 1925.
3. Official Directors Register of Pittsburgh, 1950. (ODRP)
4. The 1925 date is taken from the earliest advertising of the "Gem" carbide lamp, the Keystone Mining Catalog of 1925. Thomas Directory of Manufacturers first shows Gem manufacturing lamps in 1927. In light of this two year delay in reporting by *Thomas*, it should be realized that such dates are in many cases, approximations. Likewise, in many cases, it is the only chronological information available.
5. *Op. cit.*, ODRP. 1950.
6. *Ibid.*, 1961.

Horizontal Grier Brothers Lamp with raised lettering

THE GRIER BROTHERS COMPANY
Pittsburgh, Pennsylvania

One of the largest Pennsylvania suppliers of carbide lamps to the coal fields of that state was the Grier Brothers Company. Their first lamp was introduced in 1911,[1] and evidence indicates this lamp was probably made by the John Simmons Company, "a pinch-waist Grier Bros." variety.[2] Not long after, the Griers began to manufacturer their own products, now known to collectors as the "vertical" and "horizontal" lamps. These lamps never gained the national popularity of other major brands, but did, however, enjoy wide use in Pennsylvania.

In an attempt to diversify their product line, the Griers brought out the "Gee-Bee" lamp. While it cannot be dated with certainty, this lamp was never produced in number comparable to the regular "Grier Bros." An interesting feature of many Grier lamps was the optionally equipped Jacobson and Turja device, a novel, internal push-wire, burner-tip cleaner patented in 1921.[3]

The Grier Brothers Company, founded in 1839,

Horizontal Grier Brothers Lamps with incused lettering

THE GRIER BROTHERS COMPANY

1411-1417 Brighton Road, PITTSBURGH, PA.

Factory: PITTSBURGH, PA. Western Branch: HANCOCK, MICH.

Manufacturers of

Grier J. & T. Carbide Lamps

The Lamp You Have Waited For

Products

Carbide Lamps for Miners; Carbide Flasks—Tips and other repair parts; Miner's Dinner Pails, Powder Cans and other miner's equipment.

J. & T. Tip Cleaner

The J. & T. Tip cleaner with which Grier Lamps are now equipped is a recent invention, and thoroughly covered by patents. The cleaner is always with the lamp ready for instant use. The tip can be cleaned as often as necessary **without extinguishing the light.** It will save the miner both time and money. The cost added to the price of the lamp is very little compared to the advantages to be derived.

Construction—The device is so simple that there is nothing to get out of order from corrosion or otherwise.

Cut Showing How J. & T. Cleaners Are Installed in Grier Lamps

A small wire of special composition works inside of the lamp, being housed in a brass tube and attached to a push button. A slight pressure on the button on outside of lamp injects the wire into the gas orifice in the burner and forces out any dirt or other matter lodged in the burner.

The diameter of the cleaning wire is slightly less than the diameter of the gas opening in the burner, so that when the wire is projected into the opening the light will not be extinguished.

It is not necessary to remove the lamp from the cap.

All lamps tested before leaving our factory. Lamp equipped with Grier "Loc-on" reflectors and brass, lava insulated tips. All parts are easily replaced. Lamps made with $2\frac{1}{4}''$, $2\frac{1}{2}''$ and $3''$ reflectors. Not a better made lamp on the market.

Recent improvements made in our lamps—lower push button on tip cleaner; heavier cleaning wire, will not bend or break; stronger carbide bottoms; improved new felt holder; general construction throughout improved to make our lamps stronger and better than ever before. Lamps furnished with round or flat hooks.

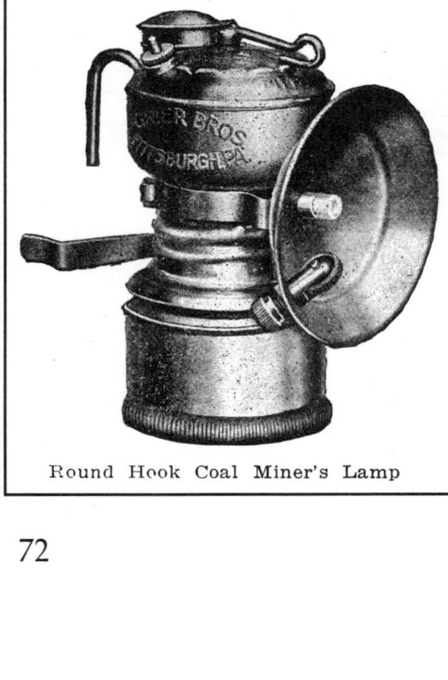

Round Hook Coal Miner's Lamp

Flat Hook Metal Miner's Lamp

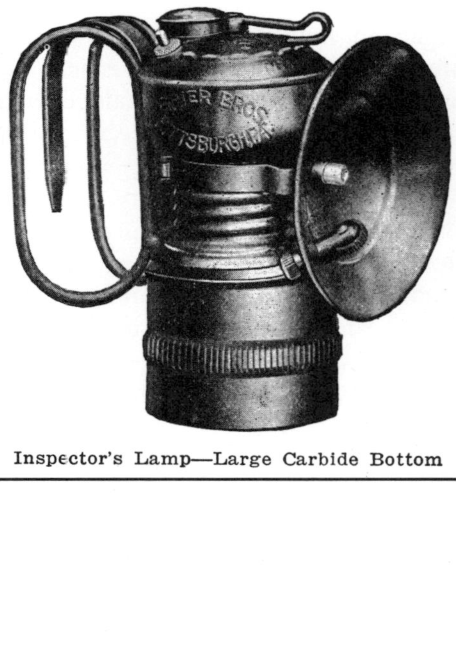

Inspector's Lamp—Large Carbide Bottom

WRITE FOR CATALOG

was located in 1909 at 124 Second Street and listed themselves as "tinware dealers."[4] This was their company address until they moved operations to 1411-17 Brighton Road in 1919-20, now trading as "Manufacturers of Carbide Miners' Lamps, Tinware, Miner's Supplies, Stove Pipe and Elbows."[5] Joseph W. Grier was President of the Company from 1909 until 1924 when the plant ceased production of the Grier brand and sold its equipment to the Gem Manufacturing Company.[6]

1. *Keystone Mining Catalog*, 1919. p. 1038.
2. This is probably another example of "private labeling" by John Simmons Company. Although proof of this is not definitive, the extreme similarity in lamp designs together with the added competition of "another brand"—albeit same manufacturer—point to Simmons as the manufacturer. The number of interesting variations in Grier Lamps, while not as common as some of the major brands, makes it difficult to place then in chronological order. Several collectors have specialized in Grier lamps and their variations almost exclusively.
3. See Letters Patent No. 1,397,170 of November 15, 1921, to Jacob W. Jacobson and John Emil Turja. Also see Reissue No. 15, 593 of May 15, 1923, to same.
4. City Directory of Pittsburgh, 1908, 1909.
5. City Directory of Pittsburgh, 1920.
6. Trade advertisement of Grier lamps after 1924 is unknown. With the Gem lamp being advertised in 1925 and incorporating the same J & T tip cleaner, it is most probable that Grier sold the rights to Gem who began making the "Gem" Lamp.

A. L. HANSEN MANUFACTURING COMPANY
Chicago, Illinois

When "Augie" Hansen left Justrite Manufacturing Company in 1920, he took with him a wealth of knowledge and design experience concerning miners' carbide lamps. It is not certain why Hansen left Justrite, but it is interesting that immediately thereafter, he formed the A. L. Hansen Manufacturing Company, located at 5037 Ravenswood Avenue, and began manufacturing the strange "Drylite" carbide cap lamp.[1]

This lamp was unusual in that the water feed was via capillarity through a system of water wicks instead of the traditional metal "dropper."[2] Not wanting to "put all his eggs into one basket," Hansen also marketed the "Hansen," a lamp with a traditional water dropper device. His "Force Feed" employed a more dynamic water valve, one that actually forced water into the carbide chamber through a positive pressure device.[3]

The company continued as a mine lamp manufacturer into the late twenties, diversfying their product line far afield of carbide lamps. Today the company survives as the A. L. Hansen Manufacturing Company, of Gurnee, Illinois, a producer of automotive parts and accessories.

1. See U.S. Trademark No. 145,684 of August 16, 1921, to A. L. Hansen. This name had been in use since September 8, 1920.

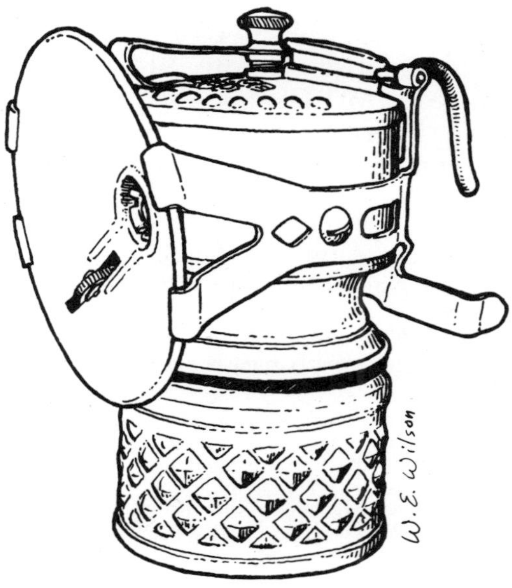

Hansen Force Feed Lamp

2. See Letters Patent No. 1,505,259 of August 19, 1924, to A. L. Hansen.
3. See U.S. Trademark No. Trademark No. 199,490 of June 9, 1925, to A. L. Hansen. This trademark had been in use since March, 1923. The portrait for the Hansen products was that of one Earl Steder, a local miner.

   See Letters Patent Nos. 1,464,988 of August 14, 1923, and 1,490,562 of April 15, 1924, both to A. L. Hansen.

MARTIN HARDSOCG COMPANY
Ottumwa, Iowa
Pittsburgh, Pennsylvania

The "Imperial" carbide mine lamp was manufactured by the Martin Hardsocg Company. The "Imperial" trademark dates back to 1898 when Hardsocg began using it on all their mining tool products.[1]

The Hardsocg cap lamp was of sturdy construction and according to one miner, would "last very well if taken care of . . . I bought [the] lights for $1.40 each in [19]30."[2]

It is not certain when the "Imperial" lamp was manufactured. The Thomas Register of Manufacturers for 1909 lists Hardsocg as a carbide mine lamp manufacturer; this listing continues until 1925.[3] Whether the company actually manufactured the lamp for these sixteen years is not known as few of the lamps have survived.

Martin Hardsocg Company was first organized in Ottumwa, Iowa, in 1871 and incorporated there in 1889. The Pittsburgh branch was organized in 1894.[4] With Martin Hardsocg as chairman and C. R. Anderson as secretary-treasurer, the company began manufacturing coal drilling machines and miners' tools, including the patented "Q" grip drills, breast augers, the patented "Black Diamond" coal and drifting picks, as well as other tools.[5]

The Pittsburgh company was located in 1909 at 920 South Avenue, Allegheny City (now known as the North Side of Pittsburgh.) By 1914, C. R. Anderson had succeeded Mr. Hardsocg as head of the company.[6] The Martin Hardsocg Company has continued since that time as a successful manufacturer of mining equipment.

1. See U.S. Trademark No. 33,224 of July 11, 1899, to the Martin Hardsocg Company of Pennsylvania. Document contains a statement attesting that the "Imperial" name has been used continuously on the company's products since January 1, 1898.
2. A. D. Grudgings, personal communication, What Cheer, Iowa, April 1, 1981.
3. *Thomas Register of Manufacturers*, 1909-1925.
4. *Allegheny County, Pennsylvania, Illustrated*. P. 198.
5. *Ibid.*, p. 198.
6. Helen M. Wilson, Librarian. The Historical Society of Western Pennsylvania. Personal Communication concerning Pittsburgh City Directories, 1914.

THE HARKER MANUFACTURING COMPANY
Cincinnati, Ohio

The Hoppe "Brite-Lite" acetylene mine lamp was first manufactured and introduced by the Harker Manufacturing Company in 1916. Sturdier and better constructed than Hoppe's earlier lamp, the "Ever-Ready," the "Brite-Lite" was constructed "of pressed steel, heavily tinned on inside and black enameled on outside, and has a nickel plated brass reflector, mounted in such a position as to make water dripping on the flame impossible."[1]

Weighing six ounces, the lamp was reported to burn for four hours on one charge of carbide. Like the "Ever-Ready," some of the "Brite-Lite" lamps had their water control valve on the side of the lamp rather than on top of the water tank. Its burner tip location was also similar to that of the "Ever-Ready" lamp.[2] A "Brite-Lite" hand lamp was also marketed on a limited basis, equipped with two folding handles as well as a hook.[3] Evidence also points to a "motorman's lamp" equipped with reflector hood and glass lens.

The Harker Manufacturing Company in 1910 was controlled by Lewis A. Harker. Established well before the turn of the century, the company's main product was carriage lamps and later, about 1905, fire extinguishers.[4]

In 1916, with Joseph A. Scallan now as owner and probably using the talents of Charles Hoppe, the Harker company entered the miners' lamp market with the "Brite-Lite" acetylene lamp. Production of the lamp continued into the late 1920s.[5] After that time, the company continued to manufacturer fire extinguishers into the 1940s.[6]

The Harker Lamp—front and rear views

1. *Engineering and Mining Journal*, July 1, 1916. p. 28.
2. *Ibid.*, October 28, 1916. pp. 792-3.
3. *The Underground Lamp Post*, Fall, 1975. p. 1.
4. Cincinnati City Directory—1891, 1898, 1905, 1907, courtesy the Cincinnati Historical Society, January 29, 1980.
5. *The Thomas Register of Manufacturers*—1928-29.
6. Cincinnati City Directory—1940-44, courtesy the Cincinnati Historical Society, January 29, 1980.

HINE-WATT MANUFACTURING COMPANY
Chicago, Illinois

Among the earliest of carbide lamps for use underground was the adapted "Columbia Model C" manufactured by the Hine-Watt Manufacturing Company. Initially in the business of manufacturing acetylene bicycle lamps, Hine-Watt fabricated a number of their "Columbia Model C" lamps with a handle instead of a bicycle handle bar attachment. At least one user as early as 1904 gave the lamp a positive review. Using his own modifications he writes

> A reflector is necessary to protect the eyes [if carried on the coat] as well as to focus the light. As perfect a water feed as can be made is the most essential item. The Columbia lamp has a very good feed an will give a very steady light for 3½ or 4 hours. It holds 3 ounces of carbide and to burn 5 hours or half shift should require 4 ounces.[1]

Sern P. Watt patented some of his ideas before the prolific Frederic E. Baldwin and deserves at least some of the credit for pioneering carbide lamps in this country.[2] Another Hine-Watt product was the belt-generator, cap burner lamp commonly referred to as the "Columbia-Jack," but not as common as the hand lamp.[3]

Hine Watt Manufacturing Company was in business before 1900 at 14-16 Canal Street. Later, they moved to 60 Wabash Street. Listed as a dealer in "auto supplies," the company was co-owned by L. A. Hine, president, and Sern P. Watt, inventor.[4] By 1916, the company had either been sold or dissolved as it was no longer listed on the Chicago Directory.[5]

1. C. E. Bailey, *Acetylene Journal*, November 1904, p. 194.
2. See Letters Patent No. 638,898 of December 12, 1899, No. 662,842 and No. 662,843 of November 27, 1900, and Design Patent No. 31,626 of October 10, 1899, all to Sern P. Watt.
3. Henry T. Pohs, *Early Underground Mine Lamps*. p. 81.
4. City Directory of Chicago—1900-1916.
5. *Ibid.*, 1916.

THE CHARLES HOPPE COMPANY
Cincinnati, Ohio

The Charles Hoppe Company manufactured one of the more unusual miners' acetylene lamps. Equipped with a peculiar, spherical water tank and an upward pointing burner tip, the "Every-Ready" lamp weighed 3 ounces, stood 4 inches high, and was 2½ inches in diameter. Made of brass, its most unusual feature was . . . "an oil-burning minature lamp which may be screwed into the top of the acetylene lamp and lighted when the other lamp is not burning at the time when the change of carbide containers is made."[1]

The "Every-Ready" with oil lamp could be purchased for $22.50 per dozen or $18.00 per dozen without oil lamp. As one contemporary noted, "The oil burning miniature or pilot strikes one as being a ridiculous refinement, somewhat like carrying a lamp in order to find one's matchbox."[2]

The company dates back to at least 1899 when they were located at 632 Main Street and listed as "manufacturers of metal goods."[3] In 1912 they moved to 121 Opera Place and there in 1913, entered the miners' carbide lamp market.[4] Charles Hoppe was listed as president, George E. Fisher was vice-president and treasurer, and Philip Gaukel worked as secretary.[5] By 1915 Charles Hoppe had apparently closed his lamp operations and had joined the Harker Manufacturing Company to develop the Hoppe "Brite-Lite" miners' carbide lamp. The "Ever-Ready" lamp was patented April 28, 1914 by Charles Hoppe, No. 1,094,902. This same patent was also applied in the manufacture of the Hoppe "Brite-Lite" in 1916.

1. "Ever-Ready Acetylene Lamp," *The Engineering & Mining Journal.*, January 3, 1914, p. 22.
2. *Ibid.*, p. 22.
3. Cincinnati City Directory—1899, courtesy of the Cincinnati Historical Society, January 29, 1980.
4. This date is based on two items; a.Hoppe's patent No. 1,094,902 was filed May 19, 1913, Serial No. 768,644; b. In the *Engineering & Mining Journal* article of January 3, 1914, it is safe to assume the reporters were writing of events at Hoppe of the past year (1913) and not of the last two days.
5. Cincinnati City Directory—1913-14, courtesy of the Cincinnati Historical Society, January 29, 1980.

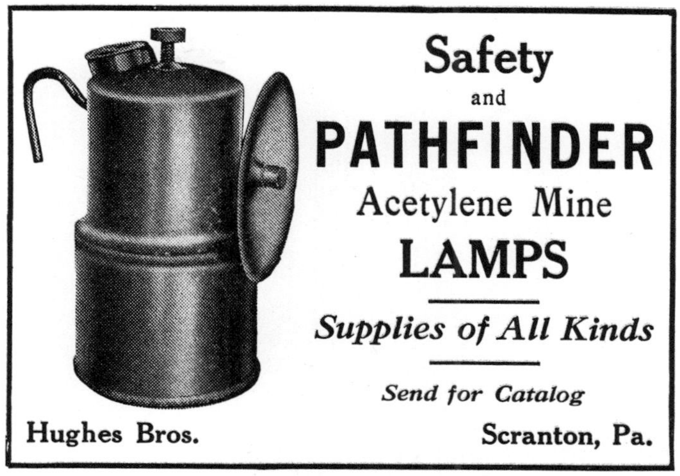

HUGHES BROTHERS
Scranton, Pennsylvania

The "Pathfinder" carbide mine lamp represents an unsuccessful attempt to enter the acetylene lamp market by an established mine safety lamp manufacturer. Founded about 1898 Hughes Brothers had gained a healthy share of the safety lamp sales in the United States by 1912.

In that year, their "Pathfinder" carbide lamp was introduced, strongly resembling the "Scranton" or "Scranto" lamp.[1] While not widely used outside of Pennsylvania, the lamp was nevertheless advertised in *Mines and Minerals* between 1912 and 1915.[2] As late as 1931 "carbide lamps" were still being advertised in the company's city directory listing.[3]

Hughes Brothers, Ralph W. and William H., started the business at 420 South Main Avenue. By 1920 Sidney R. Hughes had joined the company which was now located at 510-12 North 9th Avenue. Sidney went on to become president of the company in 1931 when it was called "Hughes Brass Works."[4] The company went out of business in 1937.

1. *Mines and Minerals Magazine*, 1912-1915.
2. For discussion on the similarity of this lamp and others to the "Scranto," etc. see the "Notes" on the Scranton Acetylene Lamp Company.
3. City Directory of Scranton, Pennsylvania, 1931.
4. *Ibid.*, 1910, 1920, 1931.

JUSTRITE MANUFACTURING COMPANY
Chicago, Illinois

The Justrite Manufacturing Company must rank as one of the world's oldest and at one time, largest carbide lamp manufacturers. Founded in 1906 and trading as "hardware specialists," Justrite under President Frederick J. Becker began exploring the promis-

Early Victor
(non-Justrite)

The Buddy

ing acetylene technology, particularly as it related to mine lighting.[1]

The first offices and shop were located at 218 Lake Street, but soon the company moved to 330-332 South Clinton Street.[2] Rapid growth and success necessitated moves to larger facilities. In 1915, the company expanded to 540 Van Buren Street and the following year, they moved to their more permanent address, 2080 Southport Avenue. Today the company continues in business at 2454 Dempster Road in nearby DesPlaines, Illinois.

Frederick Becker realized that success in the competitive carbide mine lamp market depended on a well designed, engineered lamp—one that mine superintendents and pit bosses would admire for both its ruggedness and reliability as well as its price. Only with thorough, imaginative people could Becker produce such a superior product. To his credit, he hired A. L. "Augie" Hansen, an ambitious engineer with an eye for detail and a flair for design.

Hansen, the most prolific patentee in carbide mine lamp technology, was responsible for much of Justrite's early success. His unusual "horizontal" water chamber design patent in 1912 became one of the most popular acetylene cap lamps among both ore and coal miners.[3] Always mindful of the typical miner's resistant-to-change, independent personality, Hansen also designed a carbide-spike lamp in an attempt to favorably introduce acetylene lighting to the stubborn "candle crews" still resisting the now proven, superior carbide lamp.[4] As can be seen in early advertising copy, many lamp models came with candleholders.

These efforts by Hansen were followed by ideas for lamp bottoms, lamp bottom carriers, and even exotic, swivel-reflector cap lamps which were probably never commercially available.[5] While many of Hansen's ideas were of practical use, he probably did carry his research a bit beyond the practical, everyday needs of ordinary mine illumination. For example, note the delicate, beautiful ideas embodied in his spiral, thread-feed water dropper in Letters Patent No. 1,162,915. Or consider the angled water spreader and carbide mixer in Letters Patent No. 1,224,537.

Regardless of these technological curiosities, Hansen can be credited with working out the intricacies of the flint striker and its reflector mounting.[6] Today's Justrite felt retainer and cap lamp hooks were both Hansen's ideas.[7]

By the end of World War One, Hansen sensed he could best utilize his talents by starting his own lamp company. Pooling his funds and carefully formulating his plans, he left Justrite in 1920 and formed the A. L. Hansen Manufacturing Company, manufacturers of miners' carbide lamps.

After Hansen left, William J. Frisbie continued the

technological development for Justrite. He authored the "Polygon feed" water dropper, a feature that was used continuously for the next 50 years on Justrite lamps.[8] He patented the spring-wire, thread-held cap lamp attachment,[9] and in 1921, patented the design for the "Victor" cap lamp.[10]

With the "Justrite" trademark protected by two designations and having sold hundreds of thousands of "Justrite" lamps, it is curious that the company would begin to experiment with different brand names like the "Victor" when they had a proven winner in their own "Justrite." This marketing technique was also tried with "The Buddy" carbide cap lamp.[11] Very similar to the "No. 427" Justrite, "The Buddy" did not have the distinctive horizontal water chamber.

Other new brand names were tried. "Augie" Hansen had been experimenting with cast aluminum lamps and had patented his ideas on the subject.[12] The Justrite "No. 50" was the first realization of Hansen's aluminum lamp, coming on the market around 1913-14. Known in the mines as the "Jumbo," this lamp was the predecessor to the more popular "Little Giant," first marketed in 1917. A special edition of the "Little Giant," known as the "Copper Queen" and marked as such, was made especially for those miners working this famous mine near Bisbee, Arizona. The "Little Giant" was soon accompanied by the aluminum "Uncle Sam" hook and bail lamp, a lamp known mostly for its frail, vulnerable reflector.

Still other new brands were introduced. In 1921 Justrite came out with the "Acme" half shift lamps. Available in both handle and hook 'n bail styles, the "Acme" was made of steel, with either tinned or nickel-plated finishes. These lamps were discontinued in favor of the new model Justrite "50 A" and "B" and the "56 A" and "B" about 1930.[13]

The "Victor" lamp as previously mentioned, was designed by William J. Frisbie.[14] Still willing to try other names, Justrite retooled again and the "X-Ray" was born. This cap lamp was similar to the "Victor" and differed only in a few external details. On the horizon, new plans soon led to the birth of the "Defender" cap lamp in 1922.[15] The "Defender" lamp design was also incorporated with the "X-Ray" brand name, adding to the confusion.

To mix the matter further, there is a strong, curious resemblance between the Hardsocg "Imperial," the first "X-Ray," and the obscure E. M. Haw "Fulton" cap lamp from Huntington, West Virginia. The connection between these lamps is not known conclusively and information on the E. M. Haw Company is virtually nil.

Not to be forgotten, the Justrite company continued to market its popular "Justrite" cap lamps throughout this period of name changing. With a

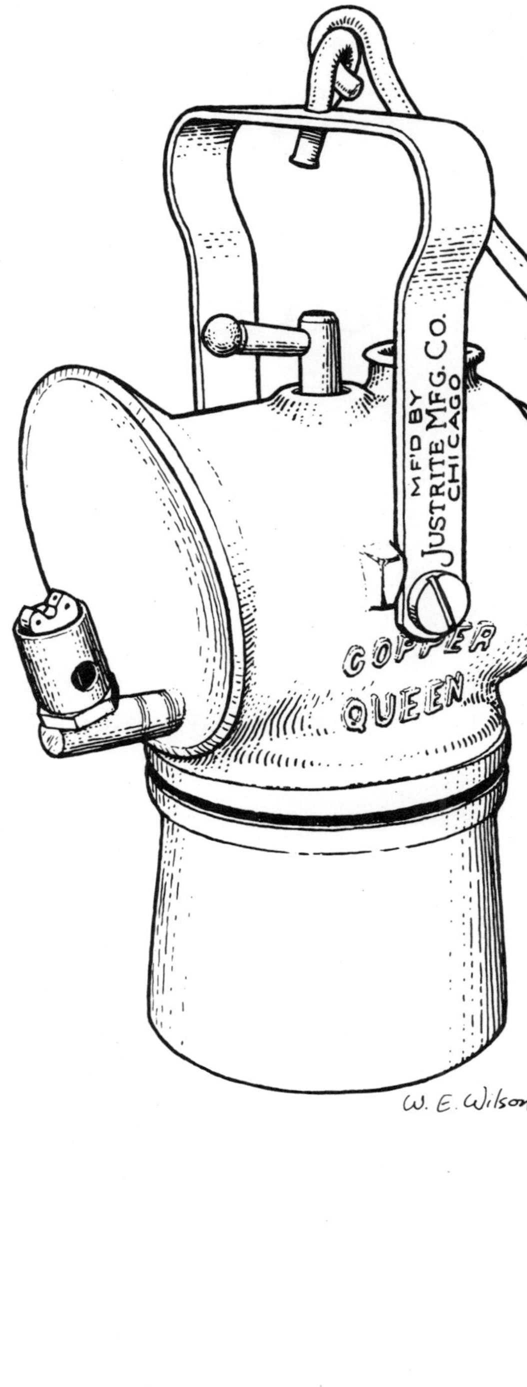

"Justrite Carbide Lamps will save you from $9.00 to $11.00 per man, per year, as compared with candles—and give more and better light"

TRADE MARK

No. 101 Lamp only—26 Gauge Brass
No. 103 Lamp only—22 Gauge Brass (Extra Heavy)

Here is the story of carbide lamp satisfaction all wrapped up in the word "Justrite." The reason is fundamental—the same as for other products of merit. It took the best material, splendid workmanship and practical workaday knowledge of ore mine conditions to produce it. There was a need for a lamp producing a steady, penetrating flame. The Justrite is just such a light. Each lamp is strongly made of brass or aluminum; equipped with self-lighter, and special water feed which eliminates clogging and produces a steady flow of gas.

Cap Lamps

Lighter Attachment
Spiral Feed

Made of polished brass—4-hour capacity — 16.7 candle-power—furnished with an extra bottom on pocket carbide can as desired. Equipped with a steel candlestick and hook—self-lighter and the "Jewel" Metal Tip.

$2\frac{5}{8}$-in. Reflector

No. 427

We make a number of models in Cap Lamps — equipped with round or flat hooks—see catalog.

3-in. Reflector

No. 597

Seamless Aluminum "LITTLE GIANT"

Spiral Feed

Burns 6 Hours

No. 110

We make many different models of Carbide Lamps. A suitable lamp for every mining purpose — send for catalog showing our complete line. Free for the asking.

Superintendent's Lamp

Nickel Plated

No. 95

Justrite Manufacturing Co.
Dept. J.
2080 Southport Ave.
Chicago - U. S. A.

wide variety of model numbers, often leading to confusion among mine superintendents, the company sometimes changed the number of a lamp listed in a previous catalog for details as trivial as a change in the lettering on the lamp.

Several very different Justrite products should be mentioned, and not to be forgotten is the "No. 10" or carbide lantern recommended to mine superintendents and railroad personnel. The "No. 44" belt-generator head lamp appeared in 1914 as Justrite's answer to the "Brilliant Search-Light."[16] The Justrite mine-car head or tail lights, "No. 240-B," and "241-B," saw use underground, but were never popular. For the mine superintendent, the Justrite "No. 95" became very popular, particularly with its nickel finish as compared to the miners' brass lamps. Indeed, from the very first carbide lamps, the nickel finish was often reserved for the Superintendent's lamp and more than a few miners held nothing but disdain for these "dainty" office illuminators.

Unlike many of its competitors, Justrite Manufacturing Company survived the "Great Depression" by diversifying products and capturing government contract work. Carbide lamps, once the principal product of Justrite, accounted for a smaller fraction of the company's total sales as the company expanded production of flashlights, fire extinguishers, shoe trees, etc. However, in 1939 the Justrite "Streamlined" cap lamp was patented,[17] and the modern day "Justrite" so familiar to sportsmen and cavers, was born. This lamp was manufactured with little change until 1973 when the company opted for the less costly plastic carbide lamps patented by Justrite engineer Frank S. Flider.[18]

Defender Style Victor

1. Chicago Business Directory—1908.
2. *Ibid.*, see also editions for 1912, 1913, 1914.
3. See Design Patent No. 42,460 of May 7, 1912, to A. L. Hansen.
4. See Design Patent No. 42,917 of August 20, 1912 to A. L. Hansen.
5. See Design Patents No. 46,280 of August 18, 1914, No. 46,905 of February 2, 1915, No. 48,109 of November 9, 1915, and No. 54,349 of December 30, 1919, all to A. L. Hansen.
6. See Letters Patent No. 1,490,562 of April 15, 1924.
7. See Letters Patent No. 1,303,658 of May 13, 1919, and No. 1,161,209 of November 23, 1915, both to A. L. Hansen.
8. See Letters Patent No. 1,407,141 of February 21, 1922, to William J. Frisbie.
9. See Letters Patent No. 1,424,463 of August 1, 1922, to William J. Frisbie.
10. See Design Patent No. 57,037 of February 1, 1921, to William J. Frisbie.
11. Collection of Larry Matthews, Hendersonville, Tennessee. Also see *Underground Lamp Post*, Henry Pohs, Editor. Fall 1977.
12. See Letters Patent No. 1,230,892 of June 26, 1917, to A. L. Hansen.
13. See Design Patent No. 82,877 of December 23, 1930, to Le Roy O. Brown.
14. Justrite was not the first manufacturer to use the "Victor" name for carbide lamps. There are in existence several non-Justrite "Victor" acetylene lamps bearing a strong resemblance to the "Scranto" style. The manufacturer of this probably older lamp is not known, but research reveals that there was a Victor Lamp Company located at Colerain and Sassafras in Cincinnati, Ohio in 1915. In 1917, the company was located at 712 Reading Road. The Thomas Register of Manufacturers listed the Victor Lamp Company as manufacturers of "acetylene and gas lamps." Also, it is of passing note that one of the principal patent attorneys handling carbide lamp patent applications was one Victor Evans. Perhaps an honorary lamp?
15. The "Defender" trademark was issued May 1, 1923 having been used since April 28, 1922. Trademark No. 167,518.

    The "Justrite" trademarks were issued repectively, No. 101,953 on January 19, 1915, dating from February 1, 1911, and No. 185,678 on June 24, 1924, dating from June 1, 1923.

    Concerning the "X-Ray" lamp, research has revealed the existence of the National X-Ray Reflector Company at 235 W. Jackson Blvd., Chicago. Owner Paul M. Hotchkin. Their registered trademark "X-Ray" was issued January 2, 1917, having been in use since March 1, 1900. Trademark No. 114,812. Interestingly enough, this trademark was issued in class 34, the same classification as the carbide lamp trademarks of Justrite, Hansen, et al. X-Ray appears to have been a manufacturer of lamp reflectors, etc.
16. *Engineering and Mining Journal*, Vol. 97, No. 26, June 27, 1914, p. 1298. The "No. 10" lantern was referred to in this article as the "No. 70."
17. See Design Patent No. 115,123 of June 6, 1939, to Charles S. Packer.
18. See Letters Patent No. 3,735,120 of May 22, 1973, and 3,953,723 of April 27, 1976, both to Frank S. Flider.

LOHMANN-HOLD MANUFACTURING COMPANY
Pittsburgh, Pennsylvania

There is very little information on the Lohmann-Hold Manufacturing Company. Regarding carbide lamps, there is a single patent, No. 1,078, 865 of November 18, 1913 to Paul Werner Lohmann. The lamp depicted in the patent graphic is similar to the odd "Maumee Duplex" lamp manufactured the next year by the Maumee Manufacturing Company, also of Pittsburgh.

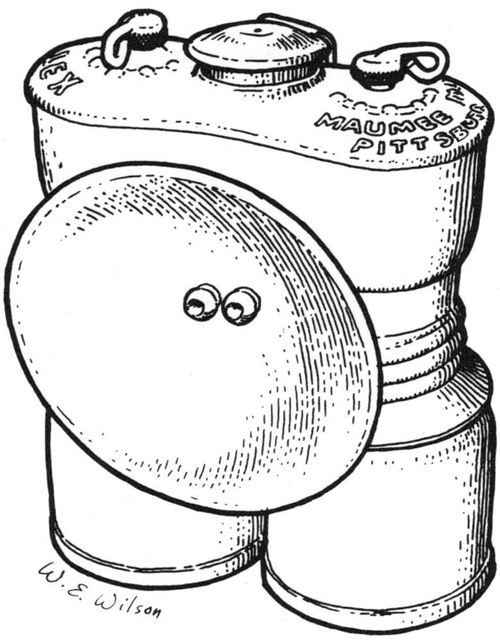

Maumee Duplex

Lohmann, interestingly enough, assigned his patent to the Lohmann-Hold Manufacturing Company on August 21, 1913, to date from the patent date of issue, November 18, 1913. Records show he received the unusually high sum of $24,500 for his rights and privileges![1]

A search of city directories failed to show the Lohmann-Hold Manufacturing Company, but a single entry for 1914 listed Otto F. and P. Werner Lohmann, mining engineers, as doing busines at 319 Third Avenue.[2] The 1915 subsequent listings do not list the Lohmanns.

A connection between the Lohmann-Hold Manufacturing Company and the Maumee lamp is possible considering the common factors of lamp design, date, and city. As Lohmann-Hold is not known to have manufactured any acetylene mine lamps, at least not under their name, any connection to the "Maumee Duplex" or any other lamp will be speculative.

1. *Digest of Assignments*, Vol. L-20. United States Trademark and Patent Office.
2. Helen M. Wilson, Librarian, The Historical Society of Western Pennsylvania. Personal communication concerning Pittsburgh city directories between 1909-1925. December 22, 1979.

THE LUNKENHEIMER COMPANY
Cincinnati, Ohio

In the last few years, several belt-generator, head-reflector acetylene lamps have been found bearing the inscription

"The Lunkenheimer Co. Cin. O. Royal No 5"[1]

As no trade advertising has been discovered, it is safe to believe that the "Lunkenheimer" lamp was an experimental or prototype lamp, probably aimed at "testing the waters" of the acetylene mine lamp market.

The lamp was apparently manufactured from a World War One artillery shell casing. It possesses a top water chamber with adjustable water-feed valve and gas pipe. A rubber hose runs from the gas pipe to the burner-reflector worn at the head. A brass belt loop is brazed to the side of the water chamber. Underneath is the carbide cup.

The Lunkenheimer Company dates back to 1862 and manufactured a wide variety of plumbing fixtures including spiggots, regrinding valves, and faucets. Many of their products were used by the mining industry; they advertised heavily in mine trade journals.

Originally known as the Lunkenheimer Brass Manufacturing Company, the company in 1921 was located at 221 East 8th Street in Cincinnati. Its massive manufacturing plant was presided over by Edmund H. Lunken, Chairman of the Board, and Shelby F. Lunken, President. Evidently, soon after the lamp was designed and tested the company decided against entering the market and further production was halted.

1. Larry Matthews, Personal communications, September 9, 1980.
2. City Directory—Cincinnati, Ohio. 1921.

MAPLE CITY MANUFACTURING COMPANY
Monmouth, Illinois

The "Maple City" acetylene miners' lamp was manufactured between 1909 and 1917 by the Maple City Manufacturing Company of Monmouth, Illinois.[1] The lamps, available both as cap and hand models, were of solid brass construction with a wire-braced reflector and highly touted "Water Flow Regulators." They were priced at $1.00 each and included a spare carbide container.[2]

The Maple City Manufacturing Company was located at 623 South First Street, just south of the C. B. & Q. Railroad tracks. In addition to carbide lamps, the company also produced an oil can that was included with every new Packard automobile. This was the company's most famous product.[3]

The Maple City address was actually a huge plant housing three companies—Maple City; Boss Manufacturing Company, maker of mittens; and Monmouth Acetylene Company. The plant was built in 1900 by the Maple City Soap Company who soon sold the property to Procter and Gamble which then leased

Maple City
Acetylene Mine Lamps

With Or Without—Which?

Will you have the Maple City Lamp with automatic feed oi water supply *plus* Water Flow Regulator?

This makes the most efficient and economical system of carbide lamp-lighting known. Or will you have it with automatic feed water supply *alone*?

This operates with lighting efficiency and carbide economy sufficient for all practical purposes.

Whichever you take you get the benefit of the strong "fool-proof" construction and the special burner which are features of Maple City Lamps.

To any Mine Superintendent upon receipt of his request on company letterhead we'll gladly send a trial lamp—*free*

To anyone else we'll gladly send a lamp prepaid upon receipt of $1.00, the usual price. This includes extra carbide container.

Maple City Bulletins will be sent for the asking to anyone interested.

Maple City Manufacturing Company

623 S. First Street **Monmouth, Illinois**

the building to other concerns. The entire plant was destroyed by fire on November 22, 1917.[4]

Advertising exists on a "Union Carbide" mine acetylene lamp. This lamp, just recently discovered, strongly resembles the unmarked Maple City design and in all likelihood was manufactured by them. Officials at Union Carbide Corporation have no record of a "Union Carbide" mine acetylene lamp, but did furnish company records concerning the "Oxweld Railroad Lamp" produced later.[5]

1. *Thomas Register of Manufacturers*, 1909.
2. *Engineering and Mining Journal*, January 6, 1912, p. 27, and July 6, 1912, p. 10.
3. Ralph B. Eckley, personal communication, January 15, 1980. Monmouth, Illinois
4. *Ibid.*, January 15, 1980.
5. Madelyn Francis, Union Carbide Corporation Information center, personal communication, January 23, 1980.

Union Carbide lamp advertisement courtesy the collection of Owen V. Fuhlrodt, Victor, Colorado via Rich Finch of Cookeville, Tennessee.

MAUMEE MANUFACTURING COMPANY
Pittsburgh, Pennsylvania

The "Maumee Duplex" is probably the strangest acetylene mine lamp encountered in this study. Manufactured only between 1914 and 1915 this odd "double lamp" probably saw only limited use in the mines.[1]

Possessing two separate carbide cups, two separate water-feed valves, and two individual burner tips, the lamp was designed to supply an auxiliary carbide light after the first carbide charge was exhausted. The water supply was designed to serve either water-feed valve.

Frank W. Law patented the "Maumee" design on May 12, 1914.[2] Two other patents, No. 1,078,865 to the Lohmann-Hold Manufacturing Company and No. 1,091,982 to Bruce B. Stone, reveal ideas and details that strongly resemble many features of the "Maumee Duplex."[3]

The Maumee Manufacturing Company was located at 7516 Bennet Street. Frank Goehring was president and Frank W. Law was secretary of the company which listed themselves as dealers in "hardware specialties."[4] In business for only two years, the company went defunct in 1915.

1. Helen M. Wilson, Librarian, The Historical Society of Western Pennsylvania, personal communication concerning Pittsburgh city directories between 1909-1925. December 22, 1979.
2. See Design Patent No. 45,758 of May 12, 1914 to Frank W. Law.
3. See Letters Patent No. 1,078,865 of November 18, 1913 to Paul Werner Lohmann and No. 1,091,982 of March 31, 1914 to Bruce B. Stone.
4. *Op. cit.*, Wilson.

ALEXANDER MILBURN COMPANY
Baltimore, Maryland

The Alexander Milburn Company manufactured a wide variety of acetylene lighting equipment and were best known for their powerful, portable searchlights. Used widely in construction, railroad and shipping yards, race tracks, circuses, industrial plants, as well as mining, this lamp could deliver up to 25,000 candlepower.[1]

With this successful product, it was a natural consequence for the Milburn company to introduce a miners' cap and hand lamp. Known as "The Milburn Miner 'A,'" the cap lamp was made of brass while the hook and bail hand lamp was of black steel construction. The hand lamp could also be employed as a cap lamp if necessary.

Although the Alexander Milburn Company was founded in 1900, it is not known when the first cap and hand lamps for miners first appeared.[2] By 1914 the Milburn company was located at 1420-1426 West Baltimore Street and headed by Alexander F. Jenkins, president. Jenkins patented several acetylene devices[3] and spearheaded the company's catalog sales, the catalog being some 48 pages long for the year 1914.[4]

By 1918 the company opened another office in New York City, at 51 East 42nd Street. With the decline in the acetylene lamp market in the 1920s, however, the Milburn company phased out acetylene illumination

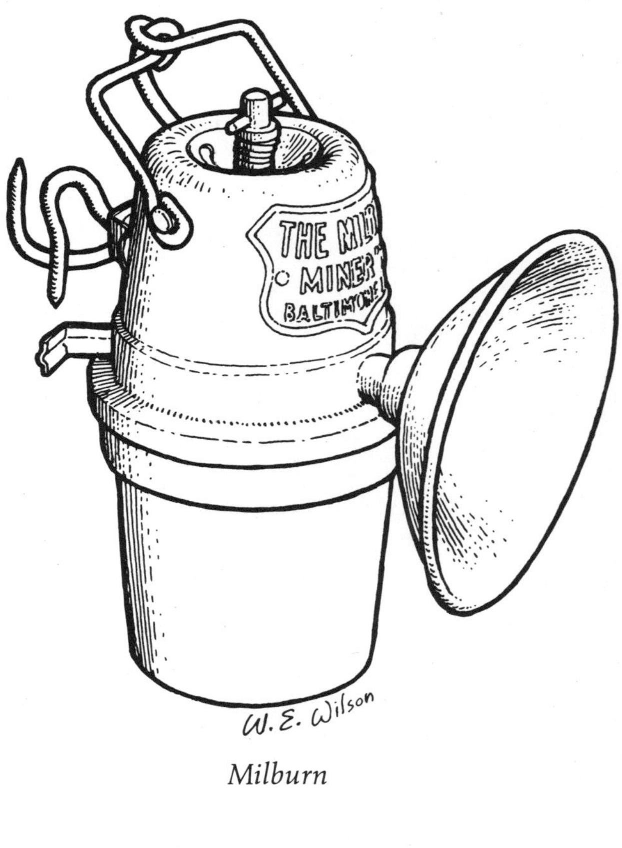

Milburn

equipment and increased manufacture of the more profitable acetylene welding and cutting apparatus.

1. *Acetylene Journal*. The Milburn Establishment, Baltimore, February, 1914, p. 310-11.
2. *Ibid.*, p. 310.
3. For a representative listing of Milburn patents see Letters Patent #1,116,822 of November 10, 1914, Nos. 1,022,356, 1,023,850 and 1,023,851 of April 23, 1912, all to Alexander F. Jenkins. Patent No. 1,023,850 was preassigned, curiously enough, to James M. Crowner of the Atlas Acetylene Lighting Company, Ottumwa, Iowa, on April 2, 1912. Only the first named patent dealt with miners' cap and hand lamps.
4. *Op. cit.*, p. 310.

   Although Alexander F. Jenkins states that the Milburn Company was formed in 1900, the "Milburn" trademark was only assigned on May 12, 1931. This trademark covered "portable acetylene lights" and was declared to have been in use since October 1, 1907. This is probably the strongest clue to when the first "Milburn" brand lamps appeared. Cap and hand lamps certainly were manufactured after this date. (See U.S. Trademark No. 283,079.)

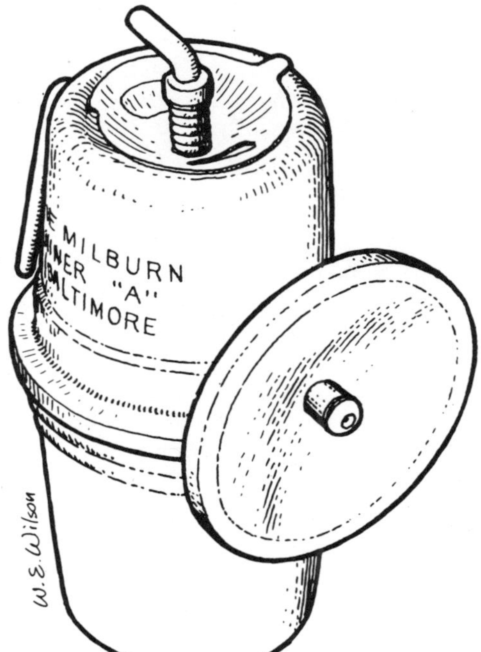

Milburn Miner "A"

NATHAN MANUFACTURING COMPANY
New York, New York

A "Nathan" acetylene miner's hand lamp has recently been found indicating yet another manufacturer's attempt to enter the acetylene lamp market. Showing some similarities to the early Baldwin and Brilliant Search Light designs, this hook and bail lamp evidently did not enjoy a wide circulation.

As early as 1876 Nathan and Dreyfus were sole manufacturers of Dreyfus' self-acting oilers and lubricators as well as Friedman's injectors for steam boilers.[1] By 1910 Nathan Manufacturing Company as the firm was now known, was headed by Alfred Nathan and located at 85 Liberty Street.[2] It is probably there that the carbide lamp venture was begun. As no trade information or advertising has come to light on the "Nathan" it is probably that the product did not enjoy a large production. Nathan continued in business as manufacturers of "injectors" through the war years. The company appears to have closed in 1922.[3]

1. New York Business Directory—1876.
2. New York Business Directory—1910-1914.
3. The last listing for Nathan Manufacturing is in 1922. Subsequent listings in New York Business Directories, give no clue to the company's fate.

Nathan

OSHKOSH METAL PRODUCTS COMPANY
Oshkosh, Wisconsin

The "Oshkosh" carbide cap lamp was produced between 1912 and 1914 by the Oshkosh Metal Products Company.[1] The cap lamp was fabricated from brass, was 2⅛ inches in diameter, and stood 2⅞ inches tall.[2] Reports on an "Oshkosh" hand lamp are still unverified.

Oshkosh Metal Products was presided over by President C. H. Hartley and Vice-president C. C.

Nelson. In January of 1912 they bought the Knippenberg Manufacturing Company, a prominent miners' oil lamp and candle holder manufacturer.[3] This acquisition and the addition of a brass foundry paved the way for the first manufacture of the brass carbide lamps by the company. By 1914, however, forced to move operations to 18 Main Street and strapped for working capital, the company ceased business.[4]

1. Mara B. Munroe, Personal communication. Oshkosh Public Library, Local history office, April 3, 1981.
2. Henry Pohs, *Underground Lamp Post*, Volume III, No. 7. Fall, 1981. p. 1.
3. Oshkosh *Daily Northwestern*, January 4, 1912, p. 8. col. 1.
4. *Op. cit.*, Munroe.
   Oshkosh City Directory, 1912, 1913, 1914, 1915.
   There is some evidence, particularly in note No. 3, that the Knippenberg firm may have made miners' acetylene lamps.

WILLIAM H. OTTO METAL GOODS COMPANY
Jersey City, New Jersey

It has recently been reported via the Underground Lamp Post that an "Ott" carbide hand lamp has been found by a collector in Ireland.[1] The lamp came to Ireland from Chile! It measures 3 inches in diameter and stands 5½ inches tall. Its 4 inch brass reflector is attached to a water chamber consisting of a brass outer sleeve and an iron or steel inner liner. The lamp bottom appears to be of iron or steel.

An inscription on the lamp reads, "William H. Ott [*sic*] Manfg. of Metal Goods, Jersey City, N. J." Research reveals a William H. Otto Metal Goods Company in Jersey City Circa 1914-1926.[2] William H. Otto had been treasurer of Thomas A. Mack Manufacturing Company in 1911, but founded his own metal goods facility shortly thereafter.[3] Located at 401 St. Paul Avenue, Otto employed his kinsman, Hugo M. Otto as treasurer. The last listing on the company was in the 1926 Jersey City Business Directory.

1. Henry Pohs, Editor, *Underground Lamp Post*. Fall, 1981. p. 4.
2. Jersey City Business Directory—1914-1926.
3. Jersey City Business Directory—1911.

SAMUEL T. SCOBY
FRANKFORT HEIGHTS, ILLINOIS

Carbide lamp collectors have known for some time of the existence of an unmarked, duplex cap lamp. While some speculated that this lamp was simply an unmarked variation of the "Maumee Duplex," others believed the lamp to have unrelated origins.

Now information has come to light showing this lamp to be the embodiment of the Samuel T. Scoby patent of 1925.[1] When Scoby, a coal miner, applied for his patent in 1922 he quickly sought and got financial backing from one Coleman Jones to manufacture and market the lamp.[2] In a similar parallel with the Samuel Simmons lamp, Scoby contracted with a St. Louis tool and machine company to fabricate the lamps. The first shipment of lamps, circa 1922-23, was subsequently found to be inferior and the remaining lamps were withdrawn from sale. This essentially ended the venture.[3]

1. See Letters Patent No. 1,567,682 of December 29, 1925 to Samuel T. Scoby.
2. Paul Kouts, personal communication, September 4, 1983.
3. In the aftermath of Scoby's lamp failure, he moved his family to Missouri in search of work. In 1926 while working in the lead mines, he was killed in a rock fall.

SCRANTON ACETYLENE LAMP COMPANY
Scranton, Pennsylvania

Among the earliest mine acetylene lamps were the various lamps manufactured by the Scranton Acetylene Lamp Company. Dating from as early as 1909,[1] the Francis H. Coffin Company began marketing the "Scranton" acetylene lamp.[2] After gaining some popularity among anthracite miners of Pennsylvania's Wyoming Valley, the name was shortened to "Scranto" in 1911 when the Scranton Acetylene Lamp Company assumed the job of manufacturing the lamp.[3] This company, with A. L. Derry Company as General Sales Agents, continued to manufacture the "Scranto" until 1916 when the American Safety

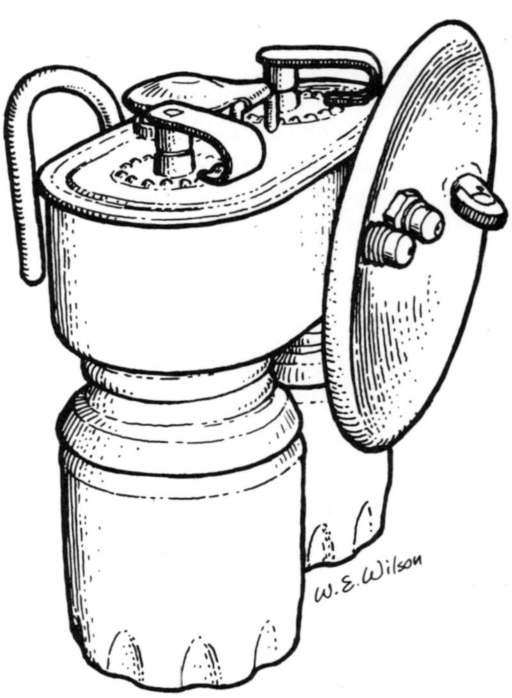

Scoby Duplex

Lamp and Mine Supply Company of Scranton bought the struggling operation. Sales of the "Scranto" ceased with this company in the early twenties.[4]

The early "Scranton" or "Scranto" acetylene mine lamp may have been the pattern for a number of similar lamps. It is known that the Abercrombie and Fitch Company of New York marketed an acetylene cap lamp under their name using the September 12, 1911 patent of David A. Williams of Scranton, Pennsylvania.[5] This lamp is much like the one patented by Llewellyn M. Evans of the Scranton Acetylene Lamp Company in 1913.[6] Hughes Brothers' "Pathfinder" lamp, also made in Scranton, strongly resembled the earlier Scranton product when it was first sold in 1912.[7] Also, the non-Justrite "Victor" lamp, manufacturer unknown, strongly resembles this family of lamps. It is possible that other different named or unnamed lamps will be found that also resemble these early examples.

1. This date is obtained from an early *Engineering & Mining Journal* advertisement dated January, 1910, The size of the ad, the early date in the year, the appearance of the lamp in a national magazine are strong evidence that the lamp existed in 1909.
2. Francis H. Coffin was listed as "General Sales Agent" for the "Scranton" lamp. It is not positive whether he was also the manufacturer, though this is probable considering his later employment with Dewar Manufacturing Company.
3. See U.S. Trademark No. 85,772 dated March 12, 1912, to Scranton Acetylene Lamp Company. Document contains the statement that the "Scranto" name was first used on September 11, 1911.
4. The date of the last production of "Scranto" lamps is not known. The American Safety Lamp and Mine Supply Company continued to sell the lamps into the early 1920s.
5. "Scranton" marked acetylene mine lamps are scarce in any collection. In view of this, the early "Scranto" lamps could have been unmarked, borne another label or name, or been marked "Scranton." As the Abercrombie and Fitch Company was an exotic supply and retail outlet for expeditions and sportsment, it is very unlikely that they were engaged in the manufacture of miners' carbide lamps. The similarity of these lamps and the common city of Scranton point to the fact that the "Abercrombie and Fitch" was probably made by the same manufacturers as the early "Scranto." See Letters Patent No. 1,002,890 of September 12, 1911, to David A. Williams.
6. See Letters Patent No. 1,081,899 of December 16, 1913, to Llewellyn M. Evans. The Scranton Acetylene Lamp Company, located at 144 Belmont Terrace, appears to have also been Evans' residence. See also, Scranton City Directories, 1913-16.
7. See advertisements in *Mines and Minerals;* 1912-1913.

See also the October 27, 1910, issue of *Mining Science,* for the article detailing a new acetylene mine lamp strongly resembling the "Scranto."

SHANKLIN MANUFACTURING COMPANY
Springfield, Illinois

Shanklin Manufacturing Company was one of the most prominent lamp manufacturers in the United States, its most famous product being the popular "Guy's Dropper" cap lamp. This lamp, with its efficient water-feed mechanism, had been designed and patented by Frank Guy, a local miner from Springfield.[1] Before Guy perfected his water-feed valve, he spent several years tinkering with carbide lamps, obtaining his first patent in 1910.[2] With limited resources and the help of the Shanklin brothers—George and Edgar—Frank Guy began manufacturing lamps on the backporch of the Shanklin home on Douglas Avenue and selling them to his fellow miners. Soon, a marketing arrangement with M. D. Barker of Springfield was struck and sales of the "Guy's Dropper" soared.[3]

By 1913 the Shanklin brothers had formed Shanklin Manufacturing Company and were gearing-up for full scale production of the lamp. As a manufacturer of "miners' supplies and metal specialities," the Shanklins now provided the "Guy's Dropper" lamp with the mass production necessary to meet a growing demand, or so they thought.

Business was so good, that in at least one case, the Shanklins "farmed out" work on the lamps to another manufacturer. Litchfield Lamp Company in nearby Litchfield, Illinois, made "Guy's Dropper" lamps around 1913-14 for Shanklin and according to one contemporary who earned $6 a week as a press operator, "We did not package the lamps, just ship[ped] the lamps to Springfield in sugar barrels."[4]

The popularity of the "Guy's Dropper" lamp and repeat business soon overwhelmed the "manufacturing facility" on the Shanklin's back porch. The lower floor of the Coats' Watch Factory was rented and Shanklins' third brother, W. Elmer, joined the operation. This location at 1020 Stanford Avenue was a vast improvement, but with continued success, the Shanklins by 1916 were ready to move into their own brick factory, a 40 x 80 foot, one story building located at 2725-29 South 11th Street.[5]

The war years of 1917-18 brought increased production and sales of carbide lamps as the nation's demand for coal and minerals climbed. In 1919 alone, 720,000 "Guy's Dropper" were produced.[6]

Besides the popular "Guy's Dropper" cap and hand lamps, Shanklin also manufactured a variation of the "Guy's Dropper" known as the "Squarelite." Patented by George Shanklin in 1916 and strongly resembling the square "Anton" lamp, the "Squarelite" came in both hand and cap models and carried the "Guy's Dropper" name.[7]

Like many of the other acetylene lamp manufacturers, the Shanklins found the post-war 1920s to bring a slow but continuous decline in lamp sales. Although the Shanklins manufactured a number of diverse articles, including miners' dinner pails, battery charging clips, and a variety of metal novelties, it was not enough to off-set this growing trend. Speculation concerning the rare "Shanklin" cap lamp suggests this to be a last ditch attempt about 1925 to relieve sagging sales.

To make matters worse, Edgar M. Shanklin was now in failing health. On the night of November 14,

(Above) The Shanklin Manufacturing Company, Springfield, Illinois.
—Courtesy Lincoln Library, Springfield, Illinois

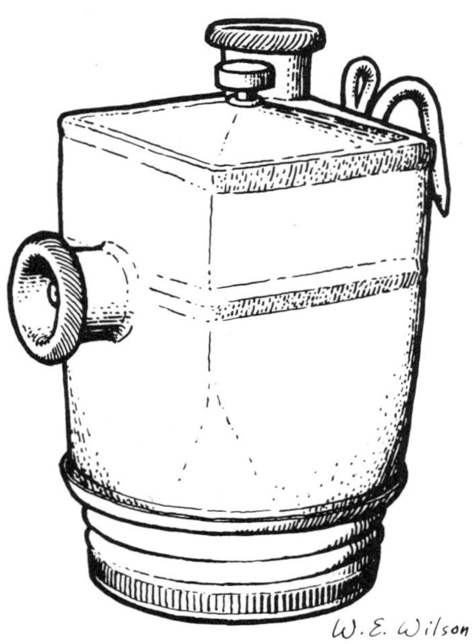

Loop-hook (Anthracite)
Squarelite-style

1926, he "walked out of the house and after eluding his nurse who accompanied him, jumped into the creek (Lick Creek, southwest of Springfield) which was swollen by recent rains."[8]

His death hastened the inevitable. In 1927 the remaining principals sold most of their interests, with W. Elmer Shanklin, however, remaining in control.

In July 1932 Shanklin Manufacturing Company was sold to Universal Lamp Company, manufacturers of the famous "Auto-Lite" cap lamp.[9] With this takeover of the Shanklin carbide lamp facility by Universal, the "Guy's Dropper" did not pass into oblivion, but instead enjoyed continued life as one of Universal's best selling products until the company discontinued manufacturing carbide lamps in 1960. Frank Guy passed the rest of his years at 2201 South Eleventh Street, witnessing the ups and downs of Shanklin Manufacturing Company. He passed away on Friday evening, February 11, 1938, at St. Johns Hospital. He was 77 years old.

1. See Letters Patent No. 1,097,716 of May 26, 1914, and No. 1,198,537 of September 19, 1916, both to Frank Guy. These are the two dates inscribed on the early cap and hand lamps, often known as the "two date" models. The "six" date models obviously came later. Examples exist of incorrect dates inscribed on the lamps. Early "one date" lamps are considered rare.
2. See Letters Patent No. 974,054 of October 25, 1910, to Frank Guy.
3. Thomas Register of Manufacturers—1912. M. D. Barker Manufacturing is listed under "acetylene lamp manufacturers" but this seems to have only been a marketing agent for Guy and the Shanklins.

SHANKLIN MANUFACTURING CO.

2725-27-29 S. Eleventh St., SPRINGFIELD, ILL.

Miners' Carbide Lamps and Supplies

Products

Miners' Carbide Lamps; Carbide Lamp Supplies.

Shanklin's Carbide Lamps

These lamps are made in a wide variety of styles, as described below.

The outstanding feature of all lamps is the Guy Patented Water Dropper, which was perfected several years ago by Mr. Frank Guy, a practical miner, and we have used it for eight years in our lamps, during which time it has proved to be a most accurate and thorough device for feeding water to our Carbide Lamps.

Guy's Patented Water Dropper

A feature of the Guy Water Dropper is the Valve Cleaner, which is a knurled thimble put on the dropper between the button and the end of the dropper tube. If dirty water is used and the dropper becomes clogged, a few turns of the valve cleaner with the fingers will open it up.

No. 22 S. L. Lamp

Has 2¾″ polished reflector, fastened to lamp with lock-nut screwed on burner tube. Reflector is strongly reinforced on back to protect it from accidents. Intended for use in rooms or entries with high roofs.

Brass lamp only, retail price $1.00.

Round or Hex. Bottom; Flat Hook or Round Hook and Brace

No. 22 Lamp, same as No. 22 S. L., without self-lighter added to reflector. Retail price $1.00.

No. 220 S.L. Lamp, same as No. 22 S. L., but nickel plated. Retail price $1.25.

No. 23 Lamp

Has large, deep, highly polished 3-inch reflector, fastened to lamp by lock-nut screwed to burner tube. Reflector strongly reinforced on back to protect it from injury.

Intended for inspectors and workmen needing an extra large light.

Round or Hex. Bottom; Flat Hook or Round Hook and Brace

Throws a bright spot of light 50 feet.

Brass lamp only. Retail price $1.00.

No. 23 S. L. Lamp has self-lighter attachment on reflector. Retail price $1.15.

No. 230 S. L. Lamp is same as No. 23 S. L., but nickelplated. Retail price $1.40.

Superintendent's Lamp

No. 10 S. L. Lamp

Superintendent's hand lamp. Can be furnished with any of our reflectors, but is usually equipped with the No. 23 S. L. reflector.

Has 50 per cent. more capacity than our regular lamp and will burn 4 hours. Has large hook and large folding handles.

Can be worn on the cap.

Brass lamp only; retail price $1.50.

No. 100 S. L. Lamp. Same as No. 10 S. L., but nickelplated. Retail price $1.75.

No. 6 S. L. Lamp

Has an extra large, deep 2¾-inch reflector highly polished. Reflector fastened to lamp by lock nut in usual manner. Reflector strongly reinforced on back to prevent injury. Intended for superintendents or inspectors needing an extra large, strong light. Throws a bright spot of light 50 feet. Lamp is same size as regular cap lamp and equipped with handles. Can be worn either on the cap or carried in the hand.

Retail price $1.35.

No. 60 S. L. Lamp. Same as No. 6 S. L., but nickelplated. Retail price $1.60.

No. 6 S. L. Lamp

No. 4 Candlestick

Slips on over bottom of lamp. Much stronger and better than a stick that clamps around the screw cap. Does away with the screw cap spreading and making a gas leak around the rubber gasket. Fits our cap lamps and four-hour lamps.

Retail price 30 cents each.

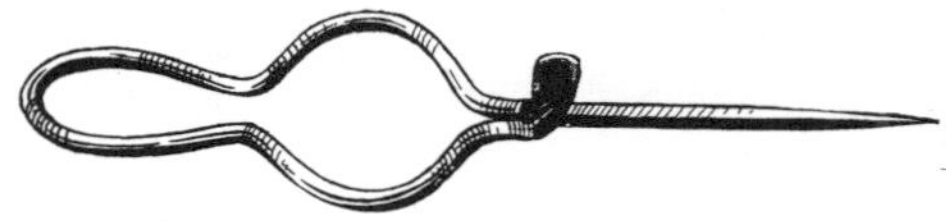

Candlestick

No. 6 Candlestick

Same as No. 4, but made to fit our half-shift lamps. Retail price 35 cents each.

Supply Kit

Contains an assortment of lamp supplies most frequently called for by the miner; also, the repair parts most likely to become lost or worn out, and which require replacement.

Conveniently arranged in a strong, durable cardboard box with hinged cover.

No. 1—$8.00; No. 2—$4.50.

Supply Kit

Personal communication with M. D. Barker's son, Paul Barker of Springfield in June 1981 confirmed the fact that his father was not a manufacturer of carbide lamps, but acted as sales agent in his mine supply business for the manufacturers, Guy and Shanklins.

4. M. K. Phillips, personal communication, March 30, 1981. Mr. Phillips states that he worked many hours on these stamping machines and "still have all my fingers." In 1981, he states "I am 84 years of age and still active." He still resides in Litchfield, Illinois and has more than a passing acquaintance with Ted Simmons, Sr., son of lamp manufacturer Samuel E. Simmons.
5. Illinois State Journal, June 18, 1930.
6. Transcripts of conversation with Mr. Joseph McGill, former Secretary-Treasurer of Shanklin Manufacturing Company. July 1, 1966. Courtesy of Richard Finch, Cookeville, Tennessee.
7. The "Squarelite" design patent was issued to George Shanklin on October 17, 1916 No. 49,782. The "Anton" lamp was patented on June 20, 1916.

   The "Guy's Dropper" trademark was No. 144,105 dated February 26, 1921, and had been in use since January, 1913.

   The "Guy's Dropper" also registered a protected label that appeared on carbide lamp boxes, etc., stating "A Lamp for Miners Invented by a Miner." Label No. 23,239 of August 21, 1921.
8. Illinois State Journal, December 22, 1926.
9. Although there may be some confusion concerning the exact circumstances and date of the Universal takeover of Shanklin, *Digest of Assignments* W-55-1, 2nd, 3rd, 4th, and W-55-2, 1st, date the sale of Shanklin in July, 1932. This is confirmed by the son of Jacob Sherman. William P. Sherman of Murfreesboro, Tennessee.

SHANKLIN METAL PRODUCTS
Springfield, Illinois

Shanklin Metal Products was formed about 1928 and was located in the old Meyer Stein Manufacturing location at 117-121 North 11th Street.[1] As a separate company from Shanklin Manufacturing Company, Shanklin Metal Products was an attempt by the Shanklins to salvage family fortunes as financial worries increased at Shanklin Manufacturing Company.

Evidence indicates that as the Shanklin brothers retired and sold their interest in the first business, the remaining brother, W. Elmer Shanklin who was president in 1927, was being pressured to step down. As a result, Elmer Shanklin, left Shanklin Manufacturing Company, founded Shanklin Metal Products and began operations in the old Meyer Stein plant.[2] His "Shanklin Metal Products" cap lamp was soon available to miners, but never really challenged the vastly popular "Guy's Dropper." By 1930 Edgar Shanklin was dead and George Shanklin had either died or retired. Shanklin Manufacturing, now under the leadership of president Theodore Hoffacker and secretary-treasurer Joseph D. McGill, continued to struggle financially. Shanklin Metal Products found the going even tougher as the Depression deepened. When Elmer Shanklin died several years later, the company closed operations. By 1934 Philip Stein was doing business at the location on 11th Street as Stein Manufacturing Company.[3] It is not known if he made

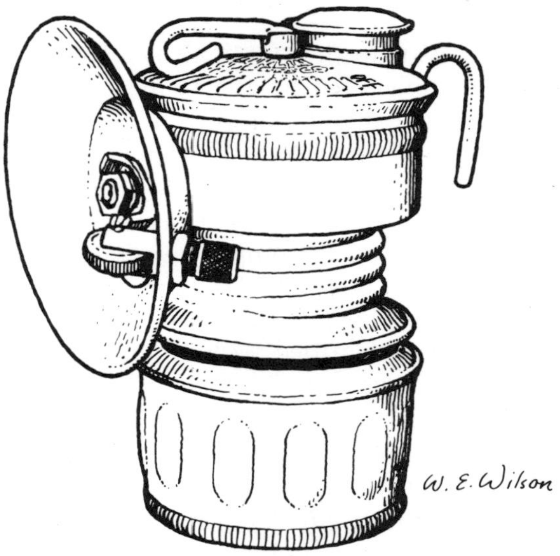

Shanklin Metal Products Co.

lamps. And of course, Shanklin Manufacturing Company had now been purchased and consolidated into Universal Lamp Company.

1. Springfield City Directory—1928.
2. It is not known whether this was a distress sale or not. Meyer Stein had been doing business here since at least 1916.
3. Springfield City Directory—1934-37.

S. E. SIMMONS MANUFACTURING COMPANY
Litchfield, Illinois

On November 22, 1910, Samuel E. Simmons, a miner and gifted mechanic, patented his ideas and improvements on miners' carbide lamps.[1] Born in Litchfield on December 3, 1867, Simmons was a natural tinkerer, possessing a strong mechanical aptitude coupled with a flair for invention and innovation.

Through his work as a coal miner, Simmons realized that an acetylene lamp would be far superior to candles and the sooty oil-wick lamp then in use. As early as 1909 he had made a working prototype of his lamp in his own home workshop.[2]

The "S. E. Simmons" lamp was approximately 3¼ inches tall and made entirely of brass. Members of the family seem to remember some of the lamps as being made of copper. The lamp was accepted locally by some of the miners and tried in the coal mines. To complement the lamp, Simmons also designed and manufactured a screw-top carbide can.[3]

When Samuel Simmons first developed his lamp, he was living at 430 South Jackson Street. In 1912 he moved his family to 802 Randall Street. There the S. E. Simmons Manufacturing Company was headquartered. It was during this time that Simmons contracted a St. Louis firm to manufacture the lamps on a

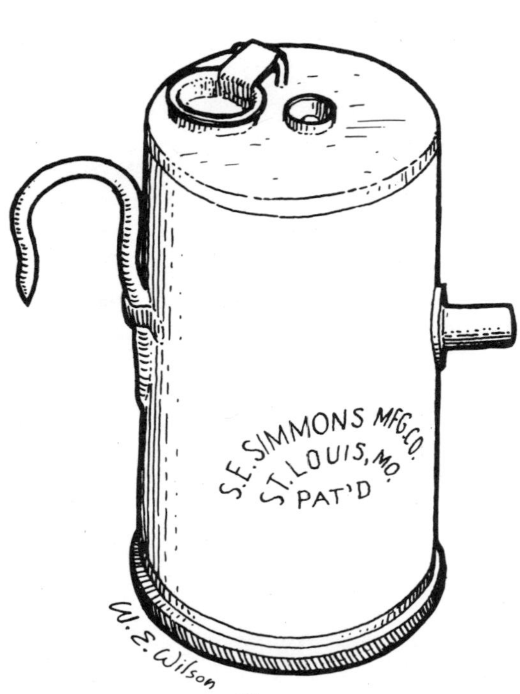

Simmons

larger scale. After an unknown number of lamps were made, a disagreement between Simmons and the company ended this arrangement.[4]

In 1916 Simmons moved again, to 202 East Martin Street and seems to have lost interest in continuing any further large scale attempts to market his carbide lamp. He did continue to produce a few lamps occasionally, giving them to his mining friends and acquaintances. He quit working in the mines in the early 1930s and on February 13, 1943, he died at the age of 75.

1. See Letters Patent No. 976,611 of November 22, 1910, to Samuel E. Simmons.
2. From evidence in the Letters Patent (see above) and family correspondence, it is probably that he first had a working lamp sometime late in 1909.
3. Much of the information, both personal and concerning the lamp, has graciously been provided by the sole surviving son of Samuel E. Simmons, Mr. Ted Simmons, Sr. Personal communications: February 22, 1981, March 9, 1981, and March 19, 1981.
4. The name of the contracting firm that manufactured the "S. E. Simmons" lamp in St. Louis is not known with certainty. Family members cannot recall the name at this late date. It is possible that the HandlanBuck Manufacturing company at 212 North 3rd Street in St. Louis, Missouri, a "miners' lamp manufacturer" listed in the 1921 Thomas Register of Manufacturers, may have indeed made these lamps.

THE SNELL LAMP COMPANY (?)
Tice, Illinois (?)

Very little information has come to light regarding "The Snell Lamp." Collectors have only a few lamp bottoms to confirm the existence of such a lamp. One such bottom, marked "The Snell Lamp, Pat. Mch. 28-11"[1] provides a clue to further research. On that date, one Frank Bradoc and William Frost patented claims for improvements in "acetylene gas lamps."[2] Interestingly enough, they assigned their patent to one Arphad Snell of Tice, Illinois. No further information is known on Snell or whether he even formed a formal company to pursue the venture. Until further lamps or information come forth, a single patent and a very few crude lamp bottoms are all that remain of Snell's "acetylene gas lamp."

1. Wendell E. Wilson, *The Nearly Complete Bottom-Guide for Carbide Cap Lamps*. Second Edition, 1982. p. 17.
2. See Letters Patent No. 988,166 of March 28, 1911, to Frank Bradoc and William Frost.

SPRINGFIELD LAMP COMPANY
(MEYER STEIN MANUFACTURING COMPANY)
Springfield, Illinois

On December 9, 1913, Meyer Stein, a machinist from Springfield, Illinois, patented a design on carbide lamps.[1] Realizing that he had a readily available market right at home, he began with Philip Stein to manufacture and market his "Springfield" acetylene cap lamp.

Located at 117-121 North 11th Street, Stein's business soon grew into "manufacturers of carbide mine lamps, mining tools and supplies."[2] Further inventions and designs led to two more patents for Stein,[3] but strong competition from nearby Shanklin Manufacturing Company and their popular "Guy's Dropper" severely hampered the "Springfield's" sales figures.

As a result of the continuing trend, Stein sold his interest in carbide lamps to Ashmead Manufacturing Company in 1922 and soon joined the company as a design consultant.[4] (Just six years later, Elmer Shanklin would move into the same building on 11th Street and found Shanklin Metal Products.)

There is conjecture that the "S & S" cap lamp was a Meyer Stein lamp, named for Philip and himself. The lamp strongly resembles the Design Patent of 1913 to Meyer Stein but cannot be conclusively designated a Springfield Lamp Company product.[5] Recently, a "Stein Dropper" lamp was located, indicating that Stein's attempts to counter Shanklin included diversification of brand names. Again, this lamp, while certainly linked to Meyer Stein, may have been produced by Ashmead on an experimental basis or by Springfield early in Meyer and Philip Stein's career.[6]

1. See Design Patent No. 45,028 of December 9, 1913 to Meyer Stein.
2. Springfield City Directory, 1915.
3. See Letters Patent No. 1,106,062 of August 4, 1916 and No. 1,376,654 of May 3, 1921, both to Meyer Stein.

4. Stein sold patent No. 1,376,654 directly to Ben Williamson of Ashmead on August 15, 1922. *Digest of Assignments*—August, 1922.
5. There appear to be two names for the Steins' business. Springfield Lamp Company appears as the actual manufacturer of the lamps while, Meyer Stein Manufacturing, a larger concern, handled the production of mining tools and supplies.
6. The "Stein Dropper" lamp is conjectured to have also been a *Philip* Stein product, manufactured in the old Shanklin Metal Products building after Elmer Shanklin's family sold out. Philip Stein moved into this building, but it is not certain what was manufactured there.

UNIVERSAL LAMP COMPANY
Chicago, Illinois
Springfield, Illinois

The moving force behind the Universal Lamp Company was one Jacob S. Sherman of Staunton, Illinois, who on January 26, 1915, secured the "Auto-Lite" trademark for miners' carbide lamps.[1] Claiming use of this logo back to May 1, 1913, Sherman seems to have started his lamp business on a shoestring budget. Through Sherman's hard work, a little luck, and a well designed product, the "Auto-Lite" rose to become one of the most popular and dependable miners' carbide lamps in the United States.

In the first years of its existence, the "Auto-Lite" was produced for Sherman from parts manufactured by Chase Brass and Copper Company of Waterbury, Connecticut.[2] These were shipped to Sherman's offices in Staunton and later Chicago for distribution. After Sherman bought out the Shanklin Manufacturing Company in 1932, Universal kept the Springfield plant operating to produce the popular "Guy's Dropper" lamp.

Universal Lamp Company controlled a number of patents including one of the original Baldwin patents from John Simmons Company in 1920.[3] Adolph C. Recker's patent for the improved reflector brace, owned by the Chase concern, was used on subsequent "Auto-Lites" for the next 36 years.[4] Henry T. Sperry, also with Chase, patented two designs in 1920, one of which strongly resembles the obscure "Arrow" cap lamp.[5] Design similarities and collector observations of Shanklin and Universal bottoms found with this lamp, lead some to conclude the Universal/Chase may have indeed been the "Arrow" manufacturer.[6] Perhaps Chase began manufacturing the "Arrow" just prior to selling its lamp interests to Universal? Indeed, this lamp and the seldom seen "Pocahontas" which strongly resembles the latter-day Universal "Guy's Dropper" might have been short-lived Universal prototypes.

Jacob Sherman himself, patented the classic "Auto-Lite" design in 1927 although the lamp was already some 14 years old.[7] For collectors, the most interesting and distinguishing feature between "early" and "late Auto-Lites" is the presence of the set-screw reflector on the early models as compared to the later "Recker" reflector brace. Another improvement was James H. Horsley's "bumper grip" patent, a popular device that is used on many of today's foreign models.[8] Horsley also patented the "modern Guy's Dropper" design for Shanklin just after it was acquired by Universal. This is the so-called "Classic Guy's Dropper" design commonly found throughout the United States by collectors.[9]

At about the time the Justrite "Acme" hand lamp was introduced, Universal brought out its version, the "Big Boy." These two lamps were similar in design, but neither remotely approached the popularity and commercial success of the smaller cap lamps.

For about two years after the Shanklin takeover by Universal, the "Guy's Dropper" was available with the famous "Auto-Lite" hexagonal bottom. Other research has revealed that there may have been a special order of fifty "Auto-Lites" with copper reflectors although this is only one employee's recollection.[10]

In the years after World War Two, the "Auto-Lite" and Guy's Dropper" shared a large portion of the carbide lamp market with the also popular "Justrite." Known as the "Big Three" by collectors, these brands were the last survivors of a once prosperous industry. With Universal's exit from the carbide lamp market in April 1960, only the brass "Justrite" remained as the last stand-by for sportsmen and cavers in the United States.

1. See U.S. Trademark No. 102,191 of January 26, 1915, to Jacob S. Sherman.
2. Interview with Mr. Joseph McGill, former Treasurer of Park Sherman Company and former officer of Shanklin Manufacturing Company and Universal Lamp Company. July 1, 1966, courtesy of R. C. Finch.
3. *Digest of Assignments*, December 29, 1920. This patent was No. 821,580 to Frederic Baldwin and included the reissue No. 13,542.
4. See Letters Patent No. 1,513,068 of October 28, 1924, to Adolph C. Recker.
5. See Design Patents No. 54,786 and No. 54,787 of March 23, 1920, both to Henry T. Sperry.
6. Among the more than several collectors who have speculated on who might have manufactured the "Arrow," the late George Bayles, that noted lamp collector from Cincinnati, believed the "Arrow" was produced by the Shanklins. (Personal communication via A. Mary Bayles, May 27, 1980.) Recent research by Paul Kouts hints to the "United Lamp Company" of Springfield as a possible candidate.
7. See Design Patent No. 72,075 of February 22, 1927, to Jacob S. Sherman.
8. See Letters Patent No. 1,790,709 of February 3, 1931, to James H. Horsley.
9. See Design Patent No. 90,467 of August 8, 1933, to James H. Horsley.
10. *Op. cit.*, McGill interview.

WHAT CHEER DRILL AND MINERS' TOOL COMPANY
What Cheer, Iowa

The What Cheer Tool Company first entered the mining equipment market in the late 1880s and by the beginning of the twentieth century, were doing over a million dollars in business a year. It was about this time that the company changed its name to the What Cheer Drill and Miners' Tool Company. Alexander Walker was now President and General Sales Agent; William Thompson was the new vice-president and treasurer.[1]

Walker, a prolific designer of mining machinery, had patented 17 different inventions concerning coal drilling machines and it was from the sale of these drilling machines that the company prospered. At one time, the company had sales agents covering the entire United States as well as Alaska, Canada, Mexico, and parts of South America.

It is uncertain when the company began to manufacturer the "What Cheer" carbide cap lamp. In all probability, the lamp did not appear until the late teens and then, saw only local use in the area mines. As mining declined in the area, so did business for the What Cheer Company. By the twenties, the company entered the automobile business, first selling Chevrolets, and later, Star and Durant automobiles. Unable to reverse a continuing downtrend in sales, the company closed in 1926.[2]

1. Mrs. Raymond Horn, personal communication, What Cheer, Iowa. March 3, 1981) (Mrs. Horn is president of the Brick School Museum in What Cheer which preserves several of the What Cheer carbide lamps.)
2. *Ibid.*, Mrs. Horn also states that local lore holds that a Sam Jones owner of a mining shop east of city hall once made the What Cheer lamp. This enterprise was known as the Star Manufacturing Company, but other information such as dates, production figures, etc., remain unknown.

WOLF SAFETY LAMP COMPANY OF AMERICA, INC.
Brooklyn, New York

The first "Wolf" acetylene mine lamps were made about 1903 in Zwickau, Germany, by the company of Friemann and Wolf.[1] They enjoyed moderate success among miners in Europe and even won second prize of 1500 Francs in a mine lamp competition sponsored by the Secretariat International du Carbure de Calcium.[2]

The German "Wolf" lamps were imported prior to World War One by the Wolf Safety Lamp Company of America, doing business at the Crystal Building, 47-49 West Street, New York City. As cap lamps were primarily an American innovation, the Wolf Company began introducing these lamps for sale in the United States. Indeed, as early as 1912, the Wolf Safety Lamp Company, Ltd., in England had begun advertising cap lamps in British mining journals. Wolf hand lamps, not to be forgotten, were also proclaimed, particularly for "their superior light—low maintenance cost and thorough dependability" as evidenced by the speedy process during the construction of the Rogers Pass Tunnel on the Canadian-Pacific Railroad.[3]

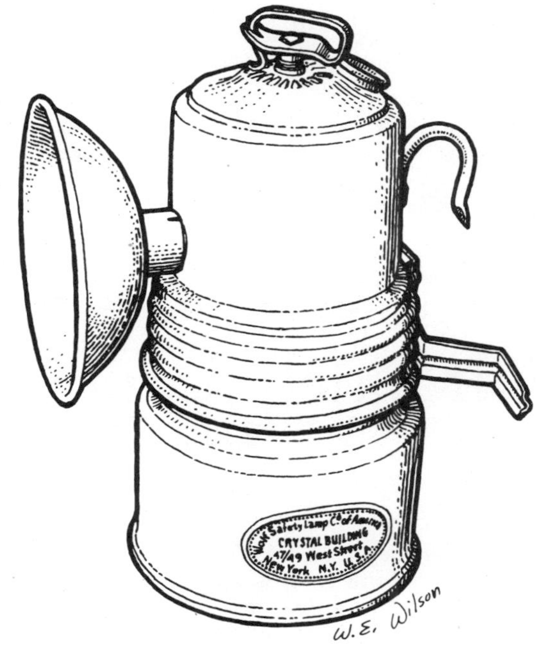

An early Wolf

In the aftermath of World War One, Domingo Anglada, the former treasurer of Wolf Safety Lamp, bought the company from the Alien Property Custodian and began manufacturing the first American "Wolf" lamps at 227 Grand Avenue in Brooklyn. Thus, as a general rule, lamps with the Brooklyn address are the post-war models, those with the New York City addresses are pre-war and of German manufacture.

With domestic production now begun, the "Wolf" lamps, particularly hand models, began to enjoy more popularity among hunters, campers, and railroad personnel. A wide variety of hand and table lamps were advertised in the 1920s and 1930s, reflecting the company's continuing favor for that design. Cap lamps did remain a large portion of Wolf's sales and to their credit, were sold on U.S. Government contract to the Civilian Conservation Corps in the 1930s. To this day, collectors can identify the "C.C.C. Wolf" lamps by their painted bottoms.[4]

Before World War One, Wolf was headed by George A. Creutzburg. The post-war president, Domingo Anglada, patented a number of ideas for Wolf including a rather odd railroad inspector's lamp. After buying Dewar Manufacturing in the late 1930s, Wolf continued to manufacturer carbide lamps, albeit on a limited scale after World War Two. In 1965, a year before his death, Anglada sold Wolf to Mine Safety Appliances (MSA) of Pittsburgh, Pennsylvania.[5]

1. *Acetylene in Wissenschaft und Industrie.*, Anleitung fur Actylen-everbraucher, March, 1916.
2. *Acetylene Journal.*, August, 1915, p. 52.
3. *Engineering and Mining Journal.*, July 1, 1916, Vol. 103, No. 1, p. 26..
4. The Wolf trademark was issued on October 28, 1924, dating back to September, 1913. See U.S. Trademark No. 190,984.
5. George Bayles, personal communication to Donald Davis, March 3, 1972. Bayles knew Anglada personally and wrote that Anglada later regretted selling out to MSA after he learned they were not interested in reviving carbide lamps for miners.

Wolf Safety Lamp Company of America, Inc.

220-224 Taaffe Place, BROOKLYN, N. Y.

Miners' Carbide Lamps

Products Made in U. S. A.

Miner's Carbide Lamps for hand, cap, trip and general utility purposes. Miners' Safety Lamps burning naphtha, vegetable oil and carbide. Tools and Accessories for all types of lamps. Newton Oiling Device for mine car wheels.

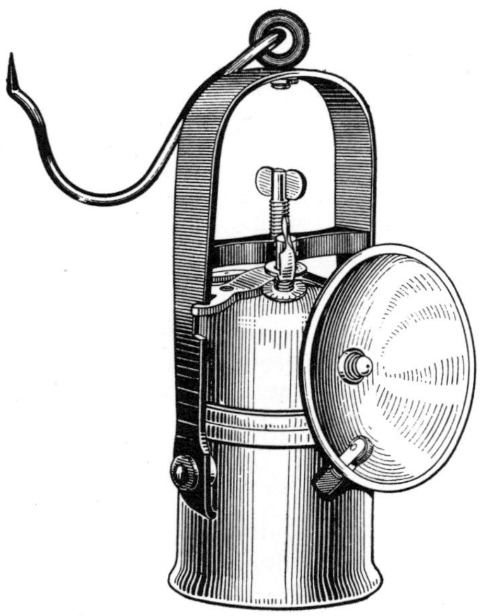

No. 905a. Hand Lamp.

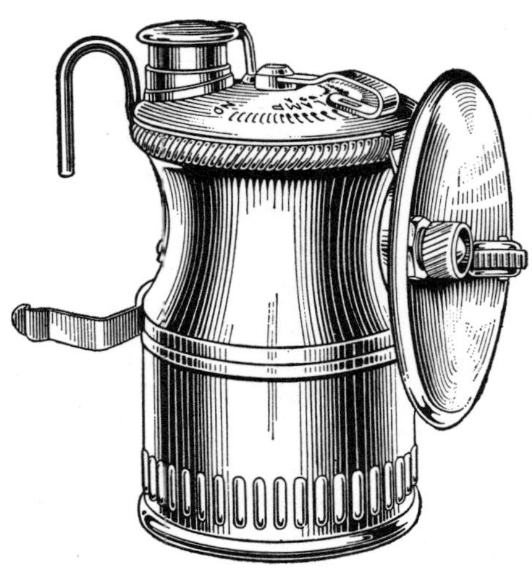

No. 911c. Cap Lamp.

No. 856. Hand Lamp.

Our Line of Lamps is undoubtedly the most complete on the market. All WOLF Lamps are guaranteed to give satisfaction or we will return your money without any question. Lamps are made in all metals and different sizes. They are either made out of steel heavily tinned or out of brass highly polished or nickel plated.

Full Description in Our General Catalogue, Copy of Which Will Be Mailed to You at Once Upon Receipt of Postal.

7

Loose Ends

There will probably never be a "complete" listing by brand name of American carbide miners' lamps, although major brands are known and many scarce lamps have been found and carefully preserved by dedicated collectors. Yet, unknown, unique lamps still surface, their origins obscure. These curious relics continue to fascinate collectors who puzzle and pick over small bits of information in hopes of making some identification, finding some new lamp, or answering some long asked question.

There are several lamps known to collectors only through trade advertising: the "Standard," "Black Diamond," "Boesch," and the "Union Carbide" to name several. Recently, one lucky collector in Tennessee reported finding a "Union Carbide" cap lamp. Perhaps some day, another collector will report finding a "Black Diamond" or some other unknown brand.

Interestingly, there are plenty of details and leads that point toward these elusive lamps. Included below are notes and facts as uncovered during the work on this book. As previously acknowledged, the author owes a great debt to the many individuals who took the time to answer questions, offer suggestions, pass on rumors, or simply say "I don't know." The following Socratic exercise will hopefully provide a "lead list" to spur those interested in further investigation.

The "Black Diamond" lamp just mentioned is one of the more elusive teasers among collectors. Known only from trade advertising, circa 1910, the Black Diamond Manufacturing Company was located at 410 Broome Street in New York City. A drawing of the lamp reveals it to be a look-alike to the "Scranto." Was this "Black Diamond" a predecessor of the "Scranto," or could it have been made by someone formerly associated with the "Scranto?" Could there have been any connection between Frederic Baldwin or John Simmons Company and the Black Diamond firm?

As noted previously, "private labeling" of carbide lamps was sometimes done by a major manufacturer for a client. This simply entailed marking a particular lamp design with the client's label, logo, or name. The lamp was then sold or distributed by the client for both profit and advertising. Justrite's "Copper Queen" mine lamp is the most noted example of this practice, but there is strong likelihood that the recently found "Abercrombie and Fitch" carbide lamp was a private label lamp. It is not likely that New York's premier expedition outfitter decided to manufacture acetylene mine lamps. The lamp itself is circa 1910 and again appears to be a "Scranto" look-alike. (Or perhaps we should say it is a "Black Diamond" look-alike.)

The "Norleigh Diamond" is another fine example of probably "private labeling." An example of this

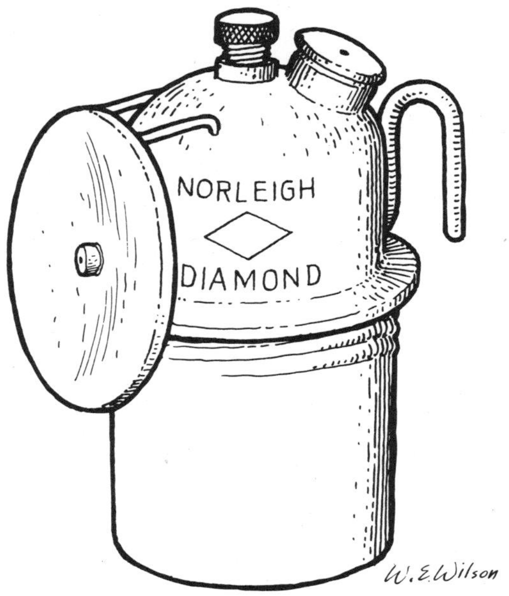

The Norleigh Diamond
After Pohs

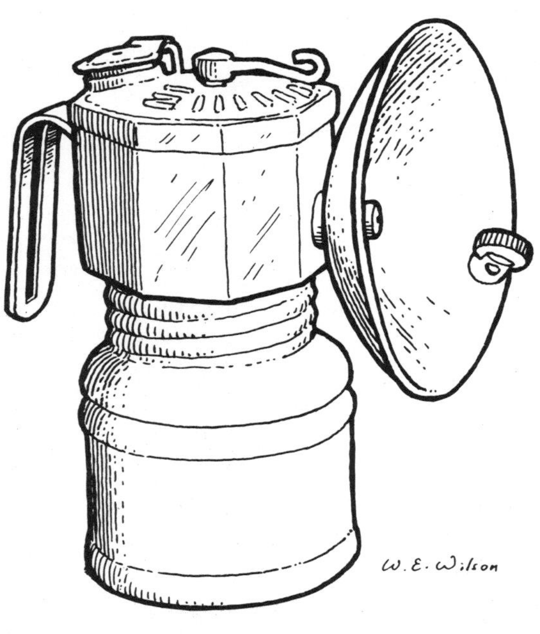

NI-BA

lamp, long rumored, has finally been acquired by collector Paul Kouts. It strongly resembles the "Maple City" design. Conjecture would indicate that the "Norleigh Diamond" name probably represented a mine, mining company, or mining supply business, a al Funk Bros.

Some of the most intriguing lamp questions deal with the many unmarked lamps that are known. In some cases, these are readily identified as a "Simmons," "Baldwin," or other prominent brand. There are, however, a number of these lamps that have defied identification. Many of these, known collectively as "anthracite" lamps in honor of the late collector and mining enthusiast George Bayles of Cincinnati, Ohio, strongly resemble the "Maple City" design. Other "anthracites" seem to have a design all their own.

The unmarked "double lamp" that had convinced some collectors into thinking they owned a scarce, unlabeled "Maumee Duplex," has recently been identified by Paul Kouts as the work of Samuel T. Scoby of Frankfort Heights, Illinois. This "Scoby" lamp was protected by Letters Patent No. 1,567,682 on December 29, 1925, and post-dates the "Maumee Duplex" by a decade.

There are several lamps of unknown manufacture that do not appear to have been "private labels." The "S & S" and "Stein Dropper" cap lamps are closely tied to the work of Meyer Stein of Springfield, Illinois, but a positive accounting of their manufacture is not documented. Did Stein make these to diversify his "Springfield" line or as prototypes for the Ashmead Manufacturing Company later?

Collectors for years have wondered who manufactured the elusive "Arrow" cap lamp. Was it Universal, Shanklin, the Chase Companies, or some more obscure firm? Is the "Pocahontas" cap lamp a sister product? Or is the "Daylight" cap lamp nothing more than an "Auto-Lite" with another name? Questions like these need answers, but there are yet other puzzles!

Who made the "Ni-Ba" cap lamp, first found in the northwest United States? Was it a foreign design? Could it have come in from the Orient? Will anyone ever discover who "Schneider" was and document the history of the "Schneider Lamp?" The rare "Denver Spike Lamp" still is a puzzle for those who seek to own this odd hybrid. And for those who own a Justrite "Victor," there is the added question of acquiring a non Justrite "Victor," another member of the "Scranto" look-alike club.

Interestingly enough, there was a Victor Lamp Company in Cincinnati that manufactured acetylene and gas lamps. In 1915 the Victor Lamp Company was located at the corner of Colerain and Sassafras, manufacturing automobile and carriage headlamps, a fact not lost with many modern-day collectors and restorers of antique automobiles. By 1917 Victor Lamp had moved to 712 Reading Road. Could they have produced these first "Victor" mine lamps? Or perhaps "Victor" was made by some unknown manufacturer, patriotically anticipating the outcome of events in war-torn Europe! Still further afield, perhaps the lamp got its name from Victor J. Evans, a prominent patent attorney who handled many of the patent applications for improvements in acetylene lamps!

One lamp of special interest is the "Fulton," supposedly manufactured by the E. M. Haw Company of Huntington, West Virginia. Yet now, there is information that the Fulton Tool Works, manufacturers of the Fulton line of mining tools, may have made this lamp. Prominently listed in the 1915 Keystone Mining Catalog as manufacturers of miners' carbide lamps, this company by 1916 had moved their business from Ohio to, of all places, Huntington, West Virginia.

A rather curious lamp, similar to the European models, was the "Sure-lite" supposedly made by the Sure-lite Lamp Company (?) of Danville, Illinois. Several examples exist, but little is known about the company or its history.

Hand-in hand with these odd and unknown lamps, there are a number of companies which are listed as manufacturers of acetylene mine lamps for which no lamp exists. As seen with the Fulton Tool Works, the Leetonia Tool Company of Leetonia, Ohio, was listed in the 1916 Keystone Mining Catalog as a carbide mine lamp producer. Will a "Leetonia" lamp ever be

Daylight
hook restored

Loop-hook "Anthracite"
Bayles Collection

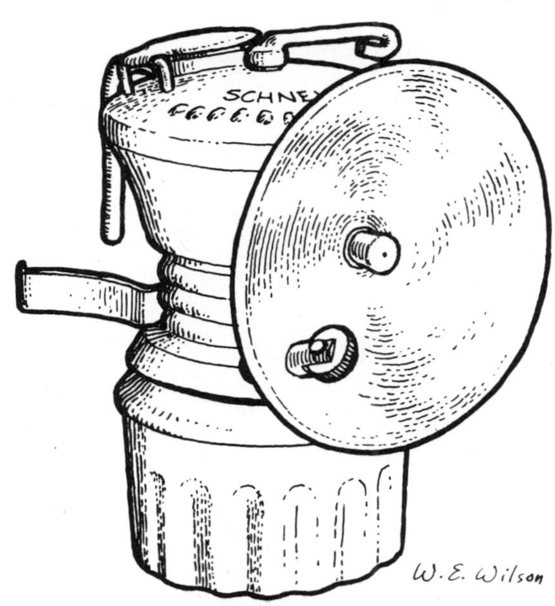

Schneider's Lamp

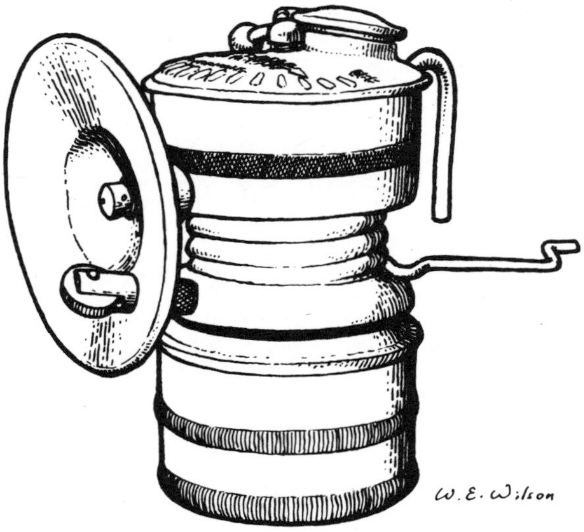

Arrow

Baldwin-style Grier Brothers
(reflector missing—probably push-on Baldwin type)

found? Was the "'Leetonia" a "private label" customer of some established lamp producer? Or perhaps, Leetonia Tool Company manufactured some of the many "anthracite" lamps that puzzle collectors.

There is likewise little known on the Macleod Company of Cincinnati which produced acetylene floor lamps and may have designed a cap lamp. Rumor of some of the bicycle lamp companies getting into the mine lamp market is substantiated by the Hine-Watt example. Can the same be said of the Badger Brass Company? It is known that the "20th Century" bicycle lamps were used for cave illumination early in the century. Were any of these lamps ever adapted for use in the mines?

One interesting lead centers on the "Universal Lamp Company" that was located at 111 5th Avenue in New York City in 1915. Although little else is known about the company, they were listed as trading in acetylene lamps. Could this "Universal Lamp Company" have had anything in common with one Jacob Sherman and another, more widely known Universal Lamp Company of a later time.

Or what about the non- Justrite "X-Ray" cap lamp? Is this a product of the National X-Ray Reflector Company of Chicago, another contemporary experimenter in acetylene lighting? Likewise, is there evidence that the Pioneer Lamp Works of 42 Hudson, New York City, had anything to do with Simmons' Pioneer line of lamps? All that is known is that Pioneer was making and selling acetylene lamps circa 1913. And as this book goes to press, collector Ralph Blanckenbeclev announces his newest find—a "Belcro No. 1."

And so the list goes . . . as fast as answers to some of these puzzles are found, other questions arise. With few surviving employees left of this once prosperous industry, only diligent research and careful collecting can clear up some of the many mysteries on the history of underground acetylene lighting.

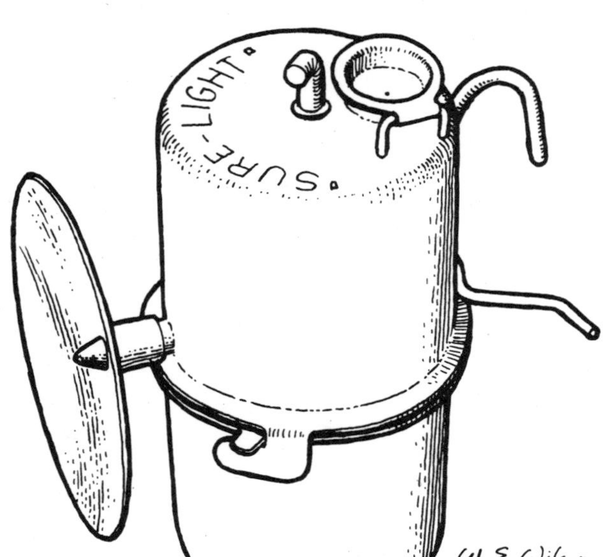

Sure-light

Belcro No. 1

8

Collecting Lamps: Caring and Coveting

If you are reading this book, you probably already collect miners' lamps or at least have more than a passing interest in the subject. As you will soon find out (if you haven't already), carbide lamp collecting is not a straight-forward pursuit, a feature that makes it a bit more interesting! Below are some hints and tips on how and where to find additions to your collection.

Antique Malls and Shops

The most obvious place is to look in antique or second-hand shops. Legendary finds still occur in such places. With relatively few people interested in the hobby compared to coin or stamp collecting, there is still plenty of opportunity to assemble a creditable collection. And who knows when a "Black Diamond" or "Standard" will turn up?

One big advantage to the collector is that to most antique dealers, a carbide lamp is a carbide lamp. They are not brand-name conscious. You'll probably pay no more for a "Milburn" cap lamp than you will for an ordinary "Justrite." Hence, it pays to know your lamps.

Larger antique malls, particularly those near old mining districts, are a favorite haunt of collectors. Here, rents and high overhead often dictate higher prices for lamps as compared to single shops, but you have the advantage of seeing the pick of many dealers without burning a lot of gas going from shop to shop. Be sure to keep your eye peeled for miners' caps, carbide cadgers, lamp repair kits, spare parts, bottoms, etc. Often these can be purchased for next to nothing if the dealer doesn't know what they were for.

Here are a couple of suggestions to ponder before entering a shop. Never wear clothing (especially t-shirts) picturing lamps or the like. This is a big tip-off to what you are after and will seriously hurt your chances when the serious haggling comes. A good practice is to first inquire about some indifferent object. Pretend to be interested in it and initiate some preliminary haggling. In the end, as you appear to be unable to agree on a satisfactory price, quietly inquire what that dirty brass, reflector-mounted object is. Show only incidental interest in it. The proprietor, still anxious to sell you, may offer very handsome terms at this point, hoping to convince you that he is a reasonable guy. Psychology is everything here!

Auctions

Rarely are carbide lamps advertised in an auction broadside. Instead, you'll have to get there plenty early and rummage around using your lamp sense to detect if there is anything of interest. Estates of farmers, miners, hunters, etc., are better bets than metropolitan or suburban sales. Yard sales are nearly always a bust unless you are near a mining area. It also seems that the older the house, the better the prospects.

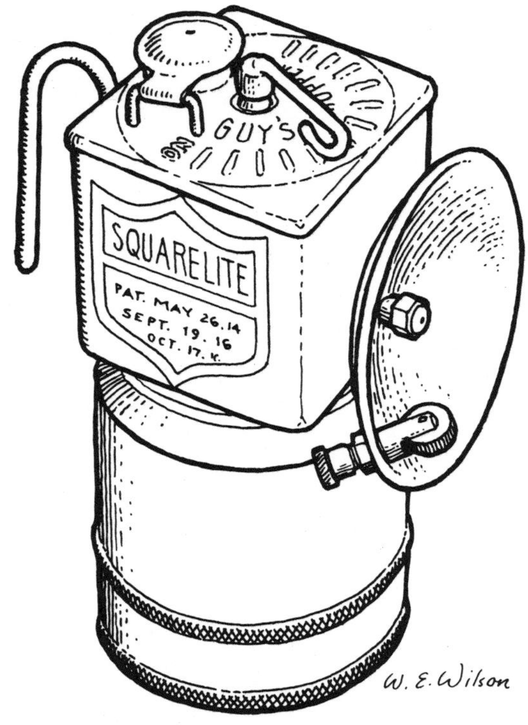

Guy's Squarelite

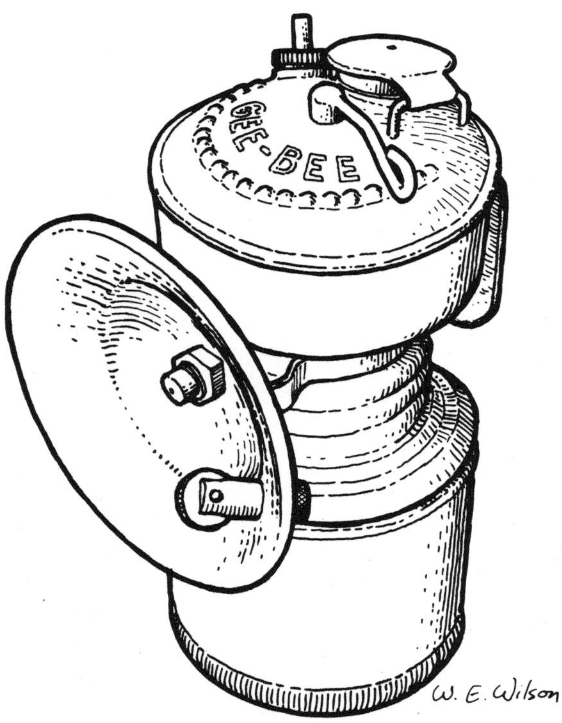

Gee-Bee

Door-to-Door

Don't spend your time on this unless you have plenty of it and need the exercise. Being in a mining area is the only reason that would keep this from being a waste of effort. Be prepared for many negative answers and some strange looks. A stop at the local hardware store, garage, or barber shop with a friendly "Howdy" for an introduction may yield some promising leads. You'll probably be dealing with old miners, widows, eccentrics, and the elderly here, so be respectful, refrain from excessive haggling and be prepared to spend an extra dollar or two. Often these folks, if you come across as respectable, will volunteer the name of another friend, relative, miner, or acquaintance who may have lamps. Be patient and listen to at least two of their old stories. If they tire of you the game is over. Don't appear over anxious to buy at first. It is amazing how that old beat up lamp in the basement will suddenly soar in value if your true lusts are detected.

Sportsmen and Cavers

It is generally a good practice to keep a couple of serviceable "Justrite" or "Auto-Lite" cap lamps around. These come in handy when one of your non-collecting friends suddenly asks you about an unusual lamp they found last week-end. You can nearly always trade a good working lamp for the older, probably nonworking model they have. In a "must have" situation, your two lamps will certainly close the deal. You can always throw in a couple of pounds of carbide to "sweeten the deal." Again, never appear over-eager to trade. They probably are aware of your hobby and will hold out for the best possible terms.

Mines and Caves

Caves are a very bad bet for lamp hunting. It is basically a fool's errand to enter a cave looking for lamps, but occasionally, cavers will report such a find. With caving rising in popularity only in the last 40 years, any lamps found in caves will certainly not be rare and surely be in bad shape.

Mines are a different story. While certainly not advocating mine exploration for lamp hunters, some nice finds have occurred in the dumps outside as well as in the depths of the excavation. Safety and property rights must take precedent before any such forays however. Generally, there are much better places to find lamps.

Old Hardware Stores and Mine Supply Houses

Although mostly a thing of the past, these businesses can still be found on occasion to have old lamps, new-in-the-box. It is hard to predict just where lamps can be found in these circumstances, but recent finds have occurred in areas where fishing is popular. One half of a case of unfired "Auto-Lites" was recently bought from a sporting goods store in Louisiana, far from ore or coal mining areas.

Other Collectors

There are plenty of places to find lamps. Sooner or later, another collector will have something that you will absolutely have to have. How you trade for that lamp is going to be your business, yet with enough imagination and persistance, you can often get what you want. You will find that many collectors prefer to trade rather than sell—some things just can't be bought.

In closing, regardless of how you pursue your hobby of lamp collecting, the "Golden Rule" is a good rule of thumb to follow in all your dealings.

Unless a lamp is unfired, it nearly always needs some form of preventative maintenance. Calcium hydroxide is a corrosive base that plays havoc with brass. Even if you choose to keep your lamp in its "original" condition, take time to neutralize the spent carbide that is slowly eating the metal in the lamp's bottom. A little vinegar left for several hours will soften and neutralize this sludge. Remove it and rinse.

By far, the best way to replace the shine on a lamp is by immersing the disassembled lamp in BRASSO. With a little "elbow grease" and patience, the lamp can be restored to a beautiful yellow tone. Avoid steel wool, screwdrivers, scrapers and anything that could cut or scratch the metal. Nickel-plated lamps respond to BRASSO or other auto chrome polishes.

NAVAL JELLY and other rust removers are sometimes used for rusty reflectors but this must be oiled or sealed to prevent a recurrence of the rust. Some of the more affluent collectors clean their lamps in ultrasonic cleaners with a variety of cleaning agents. For more on cleaning, check around with other collectors or consult Wendell Wilson's *The Early Complete Bottom-Guide for Carbide Cap Lamps*—Second Edition.

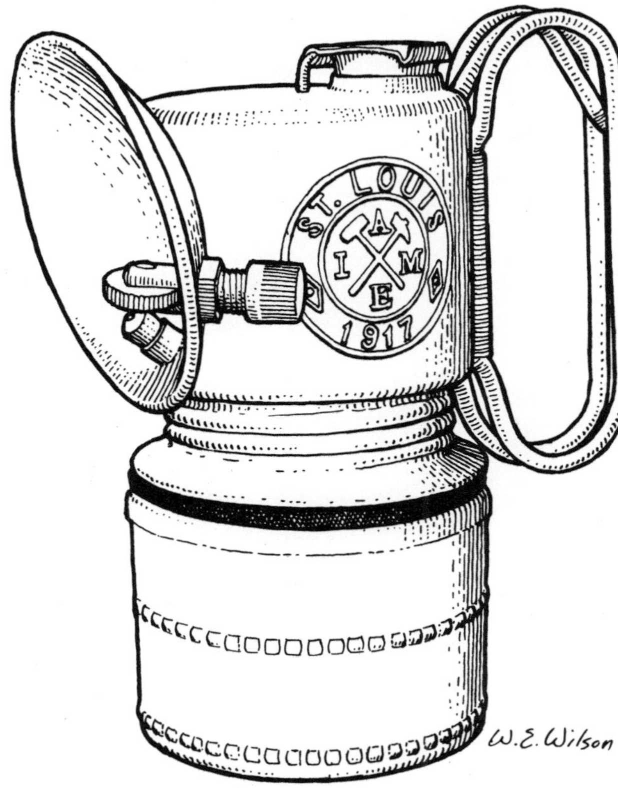

Brite-Lite
A.I.M.E. Convention

Notes to the Chapters

Chapter One

1. *Calcium Carbide and Acetylene*, George Gilbert Pond. Penn State College, State College, Pennsylvania. 1908, p. 13.
2. *Ibid.*, p. 14.
3. See Letters Patent No. 541,137 and No. 541,138 dated June 18, 1895, both to Thomas L. Willson.
4. *Op. cit.*, Pond, p. 14.

Chapter Two

1. J. D. Bowles, *Acetylene for Lighting Country Homes*. University of Missouri, March 1910.
2. See U.S. patent No. 1,22,520 dated April 10, 1917, to Matt Buzzard; also U.S. patent No. 1,805,721, dated May 19, 1931, to Frank Kogl.
3. See British patent No. 22183 of 1894, to E. Harrogate Gearing.
4. See U.S. patent No. 585,642 dated June 29, 1897, and 597,900 dated January 25,1898, both to J.C. Gallagher.
5. *The Engineering and Mining Journal.*, August 14, 1897. p. 187.
6. See British patent No. 17587 of 1898 to Leroy S. Buffington.
7. *Op. cit.*, *Engineering and Mining Journal*, November 25, 1899, p. 643.
8. *The Colliery Guardian.*, February 23, 1900, p. 354. *The Engineering and Mining Journal.*, February 17, 1900, p. 203.
9. *The Acetylene Gas Journal.*, February, 1900, p. 94.
10. *Scientific American.*, June 23, 1900.
11. See U.S. patent No. 656,874 dated August 28, 1900, to Frederic E. Baldwin.
12. See U.S. patent No. 711,871 dated October 21, 1902, to Rudolph C. Kruschke. *Acetylene Journal.*, September, 1903, p. 90.
13. See U.S. patents No. 638,898 dated December 12, 1899, No. 662,842 and 662,843 dated November 27, 1900, and Design Patent No. 31,626 dated October 10, 1899, all to Sern P. Watt, Hine-Watt Manufacturing Company, Chicago, Illinois.
14. *Op. cit.*, *Acetylene Journal*, November, 1904, p. 194.
15. See U.S. patent No. 754,472 dated March 15, 1904, to William H. Marsh.
16. See U.S. patent No. 949,349 dated February 15, 1910, to Jacob Bleser.
17. See U.S. patent No. 943,093 dated March 29, 1910, to Louis Klun.

Chapter Four

1. C. W. Owings, "Some Mine-Lighting Practices in the United States"., *Transactions of the Illuminating Engineering Society*, Vol. XXIX #1. January, 1934. p. 48.
2. *Ibid.*, p. 49. "The Electric Light at the Nunnery Colliery," *Colliery Guardian*, Vol. 40, 1880. p. 947.
3. "Safety Lamps," *Coal Age*, Vol. 1, 1911, p. 316.
4. "Electric Lighting in Mines," *Engineering and Mining Journal*, Vol. 48, 1889, p. 142.
5. *Mines and Minerals*, September 1902. p. 62.
6. Any listing of American acetylene mine lamp manufacturers before 1910 would probably be incomplete. The following companies can be included with certainty.
   1. Frederic E. Baldwin, New York, New York.
   2. Rudolph C. Kruschke, Duluth Minnesota.
   3. Maple City Manufacturing Company, Monmouth, Illinois.
   4. John S. Cummings, Tunnelton, West Virginia
   5. Justrite Manufacturing Company, Chicago, Illinois.
   6. American Safety Lamp and Mine Supply (Later Scranton Acetylene Lamp Company), Scranton, Pennsylvania. Marketed by Francis H. Coffin and Company.
   7. Hine-Watt Manufacturing Company, Chicago, Illinois.

8. Samuel E. Simmons, Litchfield, Illinois.
9. John S. Simmons Company, New York, New York.
10. Albert H. Funke, New York, New York.

Possible additions to the above list include

Hardsocg Manufacturing Company, Ottomwa, Iowa.
Knippenberg Manufacturing Company, Oshkosh, Wisconsin.
Boesch Lamp Company, San Francisco, California.

Beall Brothers Supply Company, Alton, Illinois.
Autolyte Manufacturing Company, New York, New York.
Alexander F. Milburn Company, Baltimore, Maryland.

7. James W. Paul, *Notes on Miners' Carbide Lamps.*, pp. 8-9.
8. Letter from G. M. Shoemaker to Editors of *The Colliery Engineer*, September 1913.
9. Letter from W. H. Luxton to Editors of *The Colliery Engineer*, October, 1913.
10. *Op. cit.*, Paul, p. 5.
11. *Ibid.*, p. 6.
12. Edwin M. Chance, "Portable Miners' Lamps.," *Bulletin American Institute of Mining Engineers*, No. 122, February, 1917, p. 238.
13. *Ibid.*, p. 243.
14. *Ibid.*, p. 242-243.
15. Notes of Richard Finch of conversation with Mr. Joseph McGill, former Secretary-Treasurer, Shanklin Manufacturing Company and later an employee of the Universal Lamp Company and the Park Sherman Company, July 1, 1966.
16. *Op. cit.*, Owings, p. 62.

Chapter Five

1. William J. Stephenson, "The Founding of the N.S.S.," *Journal of Spelean History;* Vol. 2, No. 4., pp. 84-6.
2. Horace C. Hovey, *The Mammoth Cave of Kentucky*, 1912. p. 122. *Acetylene Journal*, January, 1905, p. 261.
3. *Acetylene Journal*, May, 1905. p. 433.
4. *Ibid.*, September, 1908, p. 111.
5. *Ibid.*, June, 1909, p. 459.
6. *Ibid.*, August, 1914. pp. 145-6.
7. *Ibid.*, January, 1911. p. 294.
8. Bradford Cobb, owner Massanutten Caverns, personal communication, April 8, 1983.
9. Dana W. Lee, "Expedition of Carlsbad Cavern, New Mexico.," *Journal of Spelean History;* Vol. 1, No. 3.
10. George Jackson, "Russell Trall Neville Expedition in Old Salts Cave Kentucky.," *Journal of Spelean History:* Vol. 2, No. 3.
11. Chuck Young, "Improved Tip Cleaner.," *The Potomac Caver*, June 1971.
12. Tom Rice, *California Caver*, Vol. 14, No. 1, p. 8, 1962.
13. Kent Seavers, James Sturgill—personal communications, June 1977.
14. Tom Meador, "Why Not Dump it Here?," *Texas Caver*, Vol. X, No. 11, pp. 212-3, John Stade, "Calcium Carbide.," *The Underground*, Vol. VIII, No. 2, pp. 16-18.

Appendix A

UNITED STATES
CARBIDE LAMP PATENT INDEX

The Patent Index is designed to be a usable guide for those individuals desiring detailed information on the technological advances as well as the history of the miners' carbide lamp. Part One of the Index is a numerical listing in chronological order of the Letters Patent followed by the date of the patent and the name(s) of the patentee(s). Part Two is the technical index, categorizing various patents according to their respective technological improvements or characteristics. There is some overlap in Part Two due to the following reasons.

1. Patents may contain claims for more than one improvement.
2. New ideas, claims, or inventions may also fall into the Unusual category.

Part Three of the Index is a chronological listing of relevant Design Patents including date and patentee.

Copies of patents may be obtained from the United States Patent and Trademark Office for a small fee, or by visiting the Search Room of the Patent Office in person. The Search Room is presently located at 2021 Jefferson Davis Highway, Arlington, Virginia.

Part One

PATENT NUMBER	DATE	INVENTOR(S)
541,137	June 18, 1895	Thomas L. Willson
541,138	June 18, 1895	Thomas L. Willson
585,642	June 29, 1897	J. C. Gallagher
597,900	January 25, 1898	J. C. Gallagher
600,630	March 15, 1898	John Zimmerman
629,309	July 18, 1899	Henry Stearns, Sam Wilson, Ernest B. Whinnerah
629,839	August 1, 1899	Allen Winch
633,655	September 26, 1899	John A. Mosher
634,319	October 3, 1899	Charles L. Zimmerman
638,898	December 12, 1899	Sern P. Watt
639,434	December 19, 1899	Hermann Riemann
639,848	December 19, 1899	Evan Evans
640,300	January 2, 1900	F. J. Millington
649,880	May 15, 1900	August K. Stein
656,874	August 28, 1900	Frederic E. Baldwin
662,842	November 27, 1900	Sern P. Watt
662,843	November 27, 1900	Sern P. Watt
678,389	July 16, 1901	E. M. McNamara, Fred Frederick
688,926	December 17, 1901	A. L. Buffington
689,163	December 17, 1901	R. F. Carter
711,871	October 21, 1902	Rudolph C. Kruschke
731,909	June 23, 1903	Charles Johnson
738,549	September 8, 1903	Johannes A. Kruseman
754,472	March 15, 1904	William H. Marsh
821,580	May 22, 1906	Frederic E. Baldwin (Reissued March 11, 1913)
866,955	September 24, 1907	H. V. Neukirch
897,155	August 25, 1908	E. M. Rosembluth
929,296	July 27, 1909	William A. Fenton
929,454	July 27, 1909	Oliver A. Loveless
929,916	August 3, 1909	J. C. Cliver
944,789	December 28, 1909	Lester B. King
946,269	January 11, 1910	William A. Ready
949,349	February 15, 1910	Jacob Bleser
953,093	March 29, 1910	Louis Klun
953,491	March 29, 1910	Robert Seippel
956,358	April 26, 1910	Alexander F. Jenkins

956,359	April 26, 1910	Alexander F. Jenkins
960,643	June 7, 1910	Alexander F. Jenkins
969,812	September 13, 1910	Peter Togleson
974,054	October 25, 1910	Frank Guy
976,611	November 22, 1910	Samuel E. Simmons
976,785	November 22, 1910	Martin J. Curtis
986,784	March 14, 1911	John N. Tweedy
988,166	March 28, 1911	Frank Bradoc, William Frost
994,365	June 6, 1911	Frederic E. Baldwin
996,950	July 4, 1911	James Taylor
1,002,890	September 12, 1911	David A. Williams
1,005,625	October 10, 1911	Lorenzo Evans
1,007,374	October 31, 1911	A. H. Mollman
1,008,923	November 14, 1911	Martin William Moore
1,009,315	November 21, 1911	Alexander F. Jenkins
1,010,047	November 28, 1911	J. F. Heinisch
1,011,436	December 12, 1911	Alexander F. Jenkins
1,011,561	December 12, 1911	Paul Best
1,016,215	January 30, 1912	Jackson Davis
1,016,275	February 6, 1912	Henry Kerst
1,016,475	February 6, 1912	J. Corroyer
1,021,568	March 26, 1912	J. Bertha
1,022,140	April 2, 1912	Conrad Hubert
1,023,851	April 23, 1912	Alexander F. Jenkins
1,026,622	May 14, 1912	Nelson Blount
1,039,535	September 24, 1912	John M. Higgins
1,043,039	October 29, 1912	Ralph R. Arnold
1,048,870	December 31, 1912	Peter Pezzuto
1,049,437	January 7, 1913	Antony Zdanowvich
1,052,134	February 4, 1913	Ulysse Daubresse
1,053,186	February 18, 1913	John W. Larimore
1,053,516	February 18, 1913	Stillman C. Moore, Thomas Longstaff
1,055,001	March 4, 1913	A. L. Tombelaine
1,060,660	May 6, 1913	Joseph Anson
1,063,419	June 3, 1913	John Courtney
1,066,241	July 1, 1913	Frederic E. Baldwin
1,068,931	July 29, 1913	Alonzo Roach
1,070,000	August 12, 1913	Frederic E. Baldwin
1,070,352	August 12, 1913	Martin William Moore
1,071,550	August 26, 1913	Michael Cotton
1,073,980	September 23, 1913	Wolfgang Gschwendner
1,075,960	October 14, 1913	Frederic E. Baldwin
1,076,826	October 28, 1913	Augie L. Hansen
1,077,582	November 4, 1913	Frederic E. Baldwin
1,078,865	November 18, 1913	Paul Werner Lohmann
1,079,347	November 25, 1913	William F. Hudgins
1,081,503	December 16, 1913	Alexander F. Jenkins
1,081,899	December 16, 1913	Llewellyn M. Evans
1,083,427	January 6, 1914	Frederic E. Baldwin
1,084,145	January 13, 1914	Augie L. Hansen
1,087,645	February 17, 1914	Ulysse Daubresse
1,088,010	February 24, 1914	William H. Clarkson
1,088,020	February 24, 1914	Joseph Haskins
1,090,051	March 10, 1914	Paul Guilman
1,091,982	March 31, 1914	Bruce B. Stone
1,093,128	April 14, 1914	Augie L. Hansen
1,093,339	April 14, 1914	Rudolph C. Kruschke
1,094,358	April 21, 1914	Frederic E. Baldwin
1,094,902	April 28, 1914	Charles Hoppe
1,097,716	May 26, 1914	Frank Guy
1,099,908	June 16, 1914	Frederic E. Baldwin
1,101,982	June 30, 1914	Frederic E. Baldwin
1,106,062	August 4, 1914	Meyer Stein
1,109,221	September 1, 1914	Augie L. Hansen
1,113,042	October 6, 1914	Yeikichi Ohtsuka
1,113,621	October 13, 1914	Walter Harrison
1,114,208	October 20, 1914	Joseph A. Wester
1,115,157	October 27, 1914	Frederic E. Baldwin
1,116,821	November 10, 1914	Alexander F. Jenkins
1,116,822	November 10, 1914	Alexander F. Jenkins
1,119,936	December 8, 1914	John M. Downing
1,130,339	March 2, 1915	Germain Roussel
1,130,456	March 2, 1915	John H. Wininger
1,132,324	March 16, 1915	Olan V. Fortney

1,132,894	March 23, 1915	Oscar Williams, Elbert C. Baker
1,132,921	March 23, 1915	Fritz Feer
1,137,755	May 4, 1915	Augie L. Hansen
1,141,475	June 1, 1915	Arthur B. Mitchell
1,142,699	June 8, 1915	J. A. Gustafson
1,147,018	July 20, 1915	Augie L. Hansen
1,151,521	August 24, 1915	Augie L. Hansen
1,161,209	November 23, 1915	Augie L. Hansen
1,162,915	December 7, 1915	Augie L. Hansen
1,165,887	December 28, 1915	Felix H. McGurk
1,165,969	December 28, 1915	Augie L. Hansen
1,167,942	January 11, 1916	Jacob Sherman
1,168,700	January 18, 1916	L. M. Ashley, Jr.
1,174,035	March 7, 1916	Frederic E. Baldwin
1,174,036	March 7, 1916	Frederic E. Baldwin
1,174,037	March 7, 1916	Frederic E. Baldwin
1,174,112	March 7, 1916	Frederic E. Baldwin
1,178,496	April 11, 1916	Wilbur A. Cochrane
1,180,482	April 25, 1916	Augie L. Hansen
1,187,481	June 20, 1916	John B. Anton
1,187,538	June 20, 1916	Anthony Messina
1,189,014	June 27, 1916	John Sutch, Oliver Sutch, Alexander Bell
1,193,985	August 8, 1916	John M. Brock
1,194,149	August 8, 1916	John C. Davis
1,196,336	August 29, 1916	William Coburn, Charles Davis, Jules Maes
1,198,537	September 19, 1916	Frank Guy
1,198,608	September 19, 1916	Fred E. Wilkison, George J. Mitchell
1,199,267	September 26, 1916	Frank Guy
1,202,447	October 24, 1916	Henry T. Sperry
1,202,514	October 24, 1916	Augie L. Hansen
1,214,158	January 30, 1917	Augie L. Hansen
1,217,060	February 20, 1917	Joseph Potts
1,218,185	March 6, 1917	John T. Jones
1,219,980	March 20, 1917	Joseph Messina
1,224,537	May 1, 1917	Augie L. Hansen
1,225,016	May 8, 1917	James D. Cumming
1,226,237	May 15, 1917	Peter Mora
1,228,772	June 5, 1917	Augie L. Hansen
1,230,892	June 26, 1917	Augie L. Hansen
1,231,925	July 3, 1917	Guy R. McKay
1,232,572	July 10, 1917	Ronchi Lazare
1,233,663	July 17, 1917	Augie Hansen
1,237,449	August 21, 1917	Frederic E. Baldwin
1,237,450	August 21, 1917	Frederic E. Baldwin
1,237,637	August 21, 1917	Frank Guy
1,238,032	August 21, 1917	Nelson Llewellyn
1,238,729	September 4, 1917	Allen F. Adams
1,243,890	October 23, 1917	Albert T. Stewart
1,246,534	November 13, 1917	George Bowie
1,251,084	December 25, 1917	Arthur B. Mitchell
1,251,477	January 1, 1918	Wilbur A. Cochrane
1,256,770	February 19, 1918	James Brocker
1,269,047	June 11, 1918	John M. Brock
1,272,315	July 9, 1918	James H. Powers
1,274,544	August 6, 1918	Max Hendler
1,277,286	August 27, 1918	Thomas H. Bibby
1,280,214	October 1, 1918	Augie L. Hansen
1,280,849	October 8, 1918	Harry F. Roberts
1,282,913	October 29, 1918	Arthur B. Mitchell
1,283,897	November 5, 1918	Adolph C. Recker
1,289,631	December 31, 1918	John M. Brock
1,295,743	February 25, 1919	Augie L. Hansen
1,295,744	February 25, 1919	Augie L. Hansen
1,296,506	March 4, 1919	Augie L. Hansen
1,296,507	March 4, 1919	Augie L. Hansen
1,302,393	April 29, 1919	J. D. Luttrell
1,303,658	May 13, 1919	Augie L. Hansen
1,310,529	July 22, 1919	Augie L. Hansen
1,312,731	August 12, 1919	Sherman Horsley
1,319,758	October 28, 1919	John L. Conway
1,322,231	November 18, 1919	John H. Dean

1,323,618	December 2, 1919	T. M. Chance
1,324,921	December 16, 1919	William Mosso
1,328,553	January 20, 1920	Alonza D. Zimmerman
1,332,250	March 2, 1920	Henry Gall
1,334,505	March 23, 1920	George R. Shanklin
1,334,506	March 23, 1920	George R. Shanklin
1,334,570	March 23, 1920	Thomas Thompson
1,336,573	April 13, 1920	J. D. Luttrell
1,338,038	April 27, 1920	Louis E. Polhemus
1,344,867	June 29, 1920	Anthony Coyne, Lee Hash
1,349,566	August 17, 1920	William J. Frisbie
1,350,819	August 24, 1920	Mark Leonhardy
1,355,007	October 5, 1920	James A. Schooling
1,356,711	October 26, 1920	Frank Guy
1,357,595	November 2, 1920	George R. Shanklin
1,357,596	November 2, 1920	George R. Shanklin
1,358,278	November 9, 1920	Tony Bogi
1,359,330	November 16, 1920	Sam Caserto
1,359,755	November 23, 1920	George R. Shanklin
1,365,570	January 11, 1921	Fred Turner, Thomas Burton
1,366,371	January 25, 1921	Edward M. Erickson
1,368,143	February 8, 1921	Augie L. Hansen
1,371,823	March 15, 1921	Green Turner
1,373,328	March 29, 1921	Otto Harpold
1,376,654	May 3, 1921	Meyer Stein
1,385,190	July 19, 1921	Jacob S. Sherman
1,397,170	November 15, 1921	Jacob W. Jacobson, John Emil Turja (Reissued May 15, 1923)
1,397,200	November 15, 1921	Joseph Barta
1,398,673	November 29, 1921	James W. Cain
1,401,129	December 27, 1921	Ivin I. Billings
1,403,915	January 17, 1922	L. E. Polhemus
1,407,141	February 21, 1922	William J. Frisbie
1,407,142	February 21, 1922	William J. Frisbie
1,407,143	February 21, 1922	William J. Frisbie
1,407,225	February 21, 1922	Charles J. Schmidt
1,420,133	June 20, 1922	Carl H. Neusiis
1,424,463	August 1, 1922	William J. Frisbie
1,424,464	August 1, 1922	William J. Frisbie
1,429,419	September 19, 1922	William J. Frisbie
1,429,420	September 19, 1922	William J. Frisbie
1,430,247	September 26, 1922	Peter Miglio
1,431,447	October 10, 1922	William J. Frisbie
1,432,255	October 17, 1922	Russel Pearce
1,450,867	April 3, 1923	Louis E. Polhemus
1,455,130	May 15, 1923	W. W. Zickler
1,459,576	June 19, 1923	Goichi Date
1,461,572	July 10, 1923	John M. Brock
1,461,607	July 10, 1923	Newton Farley
1,464,988	August 14, 1923	John M. Brock
1,466,300	August 28, 1923	Eli Israel
1,476,962	December 11, 1923	William J. Frisbie
1,477,431	December 11, 1923	Henson Davis
1,481,079	January 15, 1924	Domingo Anglada
1,486,464	March 11, 1924	Frank Smelik
1,486,500	March 11, 1924	Samuel G. Tomasson
1,488,193	March 25, 1924	Wilbur A. Cochrane
1,490,175	April 15, 1924	Henry Luke
1,490,562	April 15, 1924	Augie L. Hansen
1,490,831	April 15, 1924	Thomas R. Jones
1,493,802	May 13, 1924	John M. Brock
1,494,721	May 20, 1924	Samuel G. Tomasson
1,498,371	June 17, 1924	Augie L. Hansen
1,498,803	June 24, 1924	William J. Mitchell
1,505,259	August 19, 1924	Augie L. Hansen
1,505,617	August 19, 1924	Archie Atkins
1,507,787	September 9, 1924	James M. Montgomery, Roscoe Ross
1,513,068	October 28, 1924	Adolph C. Recker
1,525,579	February 10, 1925	George Frost, Earnest Zacharias
1,528,848	March 10, 1925	Andrew Prader
1,531,003	March 24, 1925	Augie L. Hansen
1,552,218	September 1, 1925	Joseph Menapace
1,554,451	September 22,1925	John B. Marco
1,562,026	November 17, 1925	Charles K. Kovalchik
1,563,931	December 1, 1925	Louis Scherz
1,567,682	December 29, 1925	Samuel T. Scoby
1,569,762	January 12, 1926	Fred Lewis
1,571,554	February 2, 1926	Raymond Mull
1,578,219	March 23, 1926	Frank K. Torey
1,588,472	June 15, 1926	George S. Clark
1,605,505	November 2, 1926	John M. Brock
1,612,694	December 28, 1926	Fred R. Belt
1,612,695	December 28, 1926	Fred R. Belt
1,615,900	February 1, 1927	William J. Brown
1,617,432	February 15, 1927	Domingo Anglada
1,644,730	October 11, 1927	Augie L. Hansen
1,673,930	June 19, 1928	Domingo Anglada
1,693,463	November 27, 1928	Walter Prockter
1,712,392	May 7, 1929	Patrick McDonald
1,714,912	May 28, 1929	James M. Jones
1,716,809	June 11, 1929	Ben Williamson
1,739,916	December 17, 1929	Walter Raster
1,743,166	January 14, 1930	Charles F. Statler
1,743,235	January 14, 1930	Walter Raster
1,745,453	February 4, 1930	Harold J. Roberts
1,775,689	September 16, 1930	Paul G. Peik
1,775,690	September 16, 1930	Paul G. Peik
1,790,709	February 3, 1931	James H. Horsley
1,791,595	February 10, 1931	W. F. Hunt
1,820,642	August 25, 1931	Domingo Anglada
1,822,625	September 8, 1931	W. F. Hunt
1,852,862	April 5, 1932	Frank Smelik, Louis J. Krudwig
1,855,395	April 26, 1932	W. F. Hunt
1,863,470	June 14, 1932	Francesco Colabrese
1,925,620	September 5, 1933	Wiley B. Akers
1,943,952	January 16, 1934	W. H. Dargue
2,098,167	November 2, 1937	Charles W. Schafer
2,151,179	March 21, 1939	Domingo Anglada
2,151,471	March 21, 1939	James H. Horsley
2,214,546	September 10, 1940	Domingo Anglada
2,216,280	October 1, 1940	Charles S. Packer
2,234,033	March 4, 1941	John E. Williams
2,261,256	November 4, 1941	Casimir V. Johnson
2,293,827	August 25, 1942	Casimir V. Johnson
3,735,120	May 22, 1973	Frank S. Flider
3,953,723	April 27, 1976	Frank S. Flider

PATENT INDEX

Part Two

ACETYLENE, PROCESS OF BURNING. Referring to improved methods of burning acetylene gas in lamps.

1,296,507

ACETYLENE, SHUNTS FOR EXCESS PRESSURE. Referring to designs shunting excess acetylene to the outside of the lamp.

821,580 1,165,887 1,237,449 1,498,803

ACETYLENE, UNDERGROUND SAFETY LAMPS. Referring to safety lamps designed to employ acetylene as a fuel for illumination. (A representative listing only.)

953,491 1,011,561 1,055,001 1,174,112 1,218,185 1,323,618 1,775,689 1,775,690

ACETYLENE, LANTERNS. Referring to those patents having to do with acetylene hand-held railroad lanterns. (A representative listing only.)

1,196,336 1,228,722 1,322,231 1,791,595

ACETYLENE, WATER CHAMBER SHUNT. Referring to those designs shunting excess acetylene to the water chamber.

1,312,731 1,371,823

ALUMINUM LAMPS. Referring to those patents detailing lamps constructed of aluminum.

1,230,892 1,488,193 1,612,694 1,612,695 1,693,463

BLAST, CONCUSSION DEVICES AND IMPROVEMENTS. Referring to patents incorporating designs that prevent blast or extreme pressure changes from extinguishing the lamp's flame.

1,083,427 1,097,716 1,174,035 1,237,450 1,237,637
1,280,849

CARBIDE, AUXILIARY CHAMBERS. Referring to patents incorporating extra carbide containers or compartments into the lamp.

1,005,625 1,021,568 1,039,535 1,063,419 1,078,865
1,091,982 1,106,062 1,113,621 1,114,208 1,165,887
1,194,149 1,232,572 1,246,534 1,486,500 1,567,682
1,745,453

CARBIDE CANDLES. Referring to rare lighting devices, fueled by acetylene and resembling long candles.

639,848 754,472 1,043,049 1,336,573 1,401,129
1,420,133

CARBIDE CUP OR LAMP BOTTOMS. Referring to improvements in the carbide container design.

1,137,755 1,167,942 1,214,158

CARBIDE CUP CLEANERS. Referring to patents describing devices designed to clean spent carbide from the carbide cup.

1,106,275 1,328,553 1,432,255 1,459,576 1,494,721
1,498,371 1,507,787

CARBIDE, SUSPENDED CONTAINER. Referring to carbide cups designed to hold carbide in a wire mesh container inside the cup.

1,088,020

COMPARTMENTS, STORAGE. Referring to those patents incorporating storage containers for tips, gaskets, carbide, brushes, etc., into the lamp.

1,052,134 1,283,897 1,355,007 1,693,463 1,745,453
1,925,620

DESIGNS, EARLY. Referring to those patents illustrating early lamp designs. (A representative listing only.)

585,642 597,900 597,933 600,630 629,309
629,839 633,655 634,319 638,898 639,434
639,848 640,800 649,880 656,874 662,842
662,843 731,909 738,549 946,269 1,016,475
1,081,503

DESIGNS, GENERAL. Referring to patents detailing general improvements in carbide lamp technology.

649,880 656,874 688,926 821,580 949,349
969,812 974,054 1,002,890 1,053,516 1,097,716
1,113,042 1,115,157 1,161,209 1,165,887 1,187,481
1,202,514 1,230,892 1,238,729 1,302,393 1,303,658
1,365,570 1,366,371 1,373,328 1,430,247 1,450,867
1,612,694 1,953,723 3,735,120

Also, see British patent #17,587 for 1898.

ELECTRIC-ACETYLENE LAMP. Referring to those patents detailing a dual light supply, fueled either by electricity or acetylene.

1,358,278 1,359,330

FLINTS, IGNITER. Referring to those patents describing the use of flints in igniting the acetylene gas for illumination.

1,022,140 1,048,870 1,068,931 1,151,521 1,174,037
1,274,544 1,295,743 1,295,744 1,407,142 1,407,143
1,407,225 1,429,419 1,429,420 1,476,962 1,490,562

FLOOD LAMPS. Referring to acetylene lamps designed to light large areas underground.

956,358 956,359 960,643 1,009,315 1,011,436
1,023,851 1,116,821 1,531,003

GAS FILTERS, SCREENS, FELTS, ETC. Referring to patents detailing improvements in cleaning, filtering, or purifying acetylene generated in the lamp. Also refers to felt retaining devices and designs.

949,349 969,812 974,054 976,611 996,950
1,007,374 1,008,923 1,068,931 1,077,582 1,079,347
1,097,716 1,141,475 1,174,036 1,187,538 1,202,524
1,219,980 1,238,032 1,282,913 1,368,143 1,431,447
1,477,431 1,481,079 1,493,802 1,498,803 1,552,218
1,562,026 1,693,463 1,863,470 2,098,167

HEAT INSULATORS, GASKETS. Referring to patents describing heat protectors for hand-held lamps as well as rubber gaskets for bumper grips.

1,571,554 1,790,709 1,863,470 2,216,280

LAMP SECURING ATTACHMENTS, HANDLES. Referring to securing devices for fastening the lamp to the cap or helmet. Includes carrying handles for hand-held lamps. Also includes various improvements for securing carbide cup to water reservoir and main lamp body.

1,010,047 1,026,622 1,076,826 1,088,010 1,099,908
1,113,042 1,114,208 1,161,209 1,165,969 1,168,700
1,180,492 1,289,631 1,319,758 1,334,570 1,349,566
1,373,328 1,397,200 1,420,133 1,424,463 1,450,867
1,455,130 1,461,607 1,464,988 1,466,300 1,490,831
1,498,803 1,525,579 1,554,451 1,563,931 1,569,762
1,571,554 1,578,219 1,588,472 1,612,695 1,615,900
1,673,930 1,739,916 1,743,235 2,151,471

OIL-WATER COMBINATION LAMPS. Referring to lamps designed with dual lighting sources of either acetylene or oil, possessing both a burner tip and oil wick respectively.

953,093 1,071,550 1,073,980 1,079,347 1,094,902

ONE PIECE LAMPS. Referring to patents describing carbide lamps without the usual separation of water reservoir and carbide cup.

976,611 1,070,000 1,116,822

POCKET LAMPS. Referring to very small acetylene lamps or lamps designed to be carried conveniently in pockets, etc.

1,130,339 1,198,608

RAILROAD LAMPS. Referring to acetylene lamps designed for railroad use primarily, but occasionally used underground. (A representative listing.)

866,955 929,916 1,617,432 1,820,642 1,822,625
1,855,395 2,151,179 2,214,546 2,261,256 2,293,827

RECHARGEABLE LAMPS. Referring to lamps designed to remain lit while being recharged with carbide.

897,155 953,093 986,784 1,005,625 1,026,622
1,039,535 1,048,870 1,049,437 1,063,419 1,071,550
1,078,865 1,091,982 1,094,902 1,106,062 1,113,621
1,114,208 1,132,894 1,132,921 1,142,699 1,217,060
1,226,237 1,231,925 1,232,572 1,251,477 1,277,286
1,332,250 1,358,278 1,359,330 1,486,500 1,505,617
1,567,682 1,714,912

REFLECTORS, ATTACHMENTS, AND IMPROVEMENTS. Referring to patents detailing improvements in attaching the reflector to the body of the lamp, as well as new ideas on reflectors.

969,812 1,053,516 1,066,241 1,081,899 1,093,128
1,093,339 1,109,221 1,165,969 1,174,035 1,187,538
1,202,447 1,230,892 1,256,770 1,398,673 1,424,464
1,464,988 1,488,193 1,490,175 1,513,068 1,525,579
2,214,546

TIP CLEANERS. Referring to patents detailing tip or burner cleaning devices.

1,397,170 1,430,247 1,486,464 1,498,371 1,852,862
2,234,033 reissue 15,598

TIPS, ADJUSTABLE. Referring to burners or tips that are movable with respect to the body of the lamp.

1,116,822 1,852,862

TIPS, IMPROVEMENTS, PROTECTORS, ETC. Referring to advances in burner or tip design, particularly with respect to wind protection, fitting, etc.

929,916	994,365	1,075,960	1,083,427	1,094,358
1,147,018	1,178,496	1,187,538	1,251,084	1,269,047
1,280,214	1,280,849	1,296,506	1,338,038	1,403,915
1,461,572	1,852,862			

UPSIDEDOWN LAMP. Referring to carbide lamps designed to perform in any position, usually containing multiple generating chambers.

1,063,419	1,344,867

UNUSUAL LAMPS AND DESIGNS. Referring to unusual, different, even bizarre features patented concerning carbide lamp technology.

646,880	681,616	897,155	974,054	976,611
986,784	988,166	996,950	1,016,275	1,021,568
1,026,622	1,039,535	1,048,870	1,060,660	1,063,419
1,071,550	1,073,980	1,078,865	1,079,347	1,088,010
1,091,982	1,094,902	1,097,716	1,106,062	1,113,621
1,114,208	1,132,921	1,142,699	1,162,915	1,165,969
1,180,492	1,189,014	1,198,608	1,199,267	1,217,060
1,255,016	1,226,237	1,231,925	1,232,572	1,237,637
1,251,477	1,277,286	1,324,921	1,332,250	1,344,867
1,358,278	1,359,330	1,371,823	1,373,328	1,430,247
1,498,803	1,505,259	1,552,218	1,567,682	1,693,463
1,714,912	1,775,689	1,775,690	2,214,546	3,735,120
3.953,723				

WATER-FEED AND DROPPER IMPROVEMENTS. Referring to patents detailing improvements concerned with even water flow, control, and adjustment. Includes various designs for improved water conduits, filters, etc.

656,874	821,580	949,349	974,054	976,611
988,166	996,950	1,005,625	1,008,923	1,016,275
1,016,215	1,052,134	1,053,186	1,060,660	1,063,418
1,068,931	1,070,000	1,070,352	1,084,145	1,087,645
1,088,101	1,090,051	1,091,982	1,094,902	1,097,716
1,101,982	1,106,062	1,113,621	1,115,157	1,119,936
1,130,456	1,132,921	1,141,475	1,161,209	1,162,915
1,165,969	1,180,492	1,193,985	1,202,514	1,219,980
1,224,537	1,233,663	1,237,449	1,237,637	1,243,315
1,272,315	1,303,658	1,310,529	1,312,731	1,334,505
1,334,506	1,344,867	1,350,819	1,356,711	1,357,595
1,357,596	1,359,755	1,368,143	1,371,823	1,376,654
1,385,190	1,398,673	1,407,141	1,450,867	1,477,431
1,481,079	1,493,802	1,498,803	1,505,259	1,528,848
1,605,505	1,644,780	1,712,392	1,716,809	1,743,166
1,775,690				

WAIST GENERATORS. Referring to carbide lamps having acetylene generators at the waist of the wearer with a rubber connecting hose to the burner and reflector on the cap or helmet.

678,389	711,871	929,454	1,093,339	1,109,221
1,132,324	1,349,566			

WICK WATER-FEED. Referring to those water-feed mechanism relying on capillarity through a cloth or felt wick for lamp water-feed and control.

944,974	974,054	1,505,259

Part Three

31,626	October 10, 1899	Sern P. Watt	Hine-Watt Mfg. Co.
42,460	May 7, 1912	Augie L. Hansen	Justrite Mfg. Co.
42,917	August 20, 1912	Augie L. Hansen	Justrite Mfg. Co.
44,079	May 20, 1913	Charles Vistard	
45,028	December 9, 1913	Meyer Stein	Meyer Stein Mfg. Co.
45,514	March 31, 1914	F. E. Baldwin	
45,758	May 12, 1914	F. W. Law	Maumee Mfg. Co.
46,280	August 18, 1914	Augie L. Hansen	Justrite Mfg. Co.
46,905	February 4, 1915	Augie L. Hansen	Justrite Mfg. Co.
48,109	November 9, 1915	Augie L. Hansen	Justrite Mfg. Co.
49,782	October 17, 1916	George R. Shanklin	Shanklin Mfg. Co.
50,140	January 2, 1917	A. C. Recker	Waterbury Mfg. Co.
54,349	December 30, 1919	Augie L. Hansen	Justrite Mfg. Co.
54,786	March 23, 1920	Henry T. Sperry	Chase Companies
54,787	March 23, 1920	Henry T. Sperry	Chase Companies
57,037	February 1, 1921	William Frisbie	Justritie Mfg. Co.
63,868	February 5, 1924	Fred R. Belt	Fred R. Belt Co.
67,859	July 21, 1925	William Shanklin	Shanklin Mfg. Co.
72,075	February 22, 1927	Jacob S. Sherman	Universal Mfg. Co.
82,877	December 23, 1930	Le Roy O. Brown	Justrite Mfg. Co.
90,467	August 8, 1933	James H. Horsley	Shanklin Mfg. Co.
115,123	June 6, 1939	Charles S. Packer	Justrite Mfg. Co.

Appendix B

AMERICAN UNDERGROUND LAMPS

This listing is a compilation of research and collecting from numerous collectors across the United States. While a credit to their interest and dedication in lamp collecting, this list will probably see future additions as new, previously unlisted lamps are found.

The "rarity scale " is a reluctant addition to the list by the author in response to collectors' inquiries and urging. With "10" being the most uncommon, the lamps are numbered according to relative "brand name rarity." That is, how scarce is one particular brand name with respect to others? The numerical progression should not be interpreted in either a linear or expotential sequence, but rather, only as a rough "rule of thumb" in measuring how often a particular brand lamp is found. The number "10" refers to lamps where there are no known examples, yet where trade advertising, printed material, articles, etc., strongly pointing toward its manufacture and therefore its existence at one time. "9" is ascribed to those lamps where at least one lamp is definitely known and others suspected. The absence of any lamps with a "3" is explained by the need to create a "gulf" between the common "Guy's Dropper" (2), "Justrite" (1), and "Auto-Lite" (1), and the more scarce lamps such as "Baldwin," and "Wolf," and "Big Boy," all ranked at "4." There is no attempt or intent to infer financial value in this list and all "values" ascribed to these lamps are strictly between the seller and buyer.

Those collectors interested in researching the individual styles and designs within a particular brand name are referred to the various pamphlets in the *Miner's Carbide Lamp Reference* by Paul Kouts and the *Nearly Complete Bottom-Guide for Carbide Cap Lamps* by Wendell E. Wilson.

BRAND NAME	RARITY	MANUFACTURER
1. ABERCROMBIE & FITCH	9	Unknown; strongly resembles Scranto and Black Diamond
2. ACME	5	Justrite Manufacturing Co.
3. AMERICAN	9	American Lamp and Specialty Co.
4. "ANTHRACITE"	6	Unknown; various styles . . . included for completeness
5. ANTON	8	George B. Anton and Brother

6. ARNOLDS'S CARBIDE CANDLE	8	William Ainsworth and Sons
7. ARROW	6	Unknown; possibly United Lamp Co., Shanklin Manufacturing Co.
8. AUTO-LITE	1	Universal Lamp Company
9. BALDWIN	4	Frederick E. Baldwin, John Simmons Company
10. BIG BOY	4	Universal Lamp Company
11. BLACK DIAMOND	10	Black Diamond Manufacturing Co.
12. BOESCH	10	Boesch Lamp Company
13. BRILLIANT SEARCH LIGHT	4	Rudolph C. Kruschke; Brilliant Search Light Co.
14. BRITE-LITE	7	Charles Hoppe; most made by Harker Manufacturing Co.
15. BUDDY	5	Ashmead Manufacturing Co.
16. CHANCE	10	T. M. Chance, prototype only?
17. COLUMBIA	6	Hine-Watt Manufacturing Co.
18. COOLITE or COOLAMP	10	Coolamp Company
19. COPPER QUEEN	8	Justrite Manufacturing Co.
20. DAYLIGHT	9	Unknown; strongly resembles Auto-lite.
21. DEFENDER	6	Justrite Manufacturing Co.
22. DENVER SPIKE LAMP	9	Denver Lamp Company
23. DEW-R-LITE	5	Dewar Manufacturing Company
24. DRYLITE	9	A. L. Hansen Manufacturing Co.
25. ELKHORN	7	Ashmead Manufacturing Co.
26. EVER-READY	9	The Charles Hoppe Co.
27. FORCE-FEED	7	A. L. Hansen Manufacturing Co.
28. FULL MOON	9	A. H. Funke
29. FULTON	9	Fulton Tool Works?
30. FUNK BROS. MFG. CO.	9	Funk Brothers Manufacturing; John Simmons Co.?
31. GEE-BEE	8	Grier Brothers Manufacturing Co.
32. GEM	8	Gem Manufacturing Co.
33. GRIER BROS.	5	Grier Brothers Manufacturing Co.
34. GUY'S DROPPER	2	Shanklin Manufacturing Co., Litchfield Lamp Co., Universal Lamp Co.
35. HANSEN	7	A. L. Hansen Manufacturing Co.
36. IMPERIAL	9	Martin Hardsocg Company
37. I-T-P	4	Dewar Manufacturing Co.
38. JUSTRITE	1	Justrite Manufacturing Co.
39. JUSTRITE STREAMLINED	1	Justrite Manufacturing Co.
40. LITTLE GIANT	6	Justrite Manufacturing Co.
41. LU-MI-NUM LAMP	5	Fred R. Belt Company
42. LUNKENHEIMER	9	The Lunkenheimer Company
43. MAPLE CITY LAMP	7	Maple City Manufacturing Co.
44. MAUMEE-DUPLEX	8	Maumee Manufacturing Co.
45. MILBURN MINERS' A	8	Alexander Milburn Co.
46. NATHAN	9	Nathan Manufacturing Co.
47. NI-BA	9	Unknown
48. NORLEIGH DIAMOND	9	Unknown
49. OSHKOSH METAL PRODUCTS CO.	9	Oshkosh Metal Products Co.
50. OTTO (OTT)?	9	William H. Otto Metal Goods Co.
51. PATHFINDER	7	Hughes Brothers

52. POCAHONTAS	8	Unknown; speculation centers on Shanklin brothers
53. RED STAR	10	Beall Brothers Supply Co.
54. S & S	9	Unknown; strong evidence of Meyer Stein involvement
55. SCHNEIDER'S LAMP	9	Unknown
56. SCOBY LAMP	9	Samuel Scoby
57. SCRANTO	8	Scranton Acetylene Lamp Co.
58. SCRANTON	9	Scranton Acetylene Lamp Co.
59. SHANKLIN	8	Shanklin Manufacturing Co.
60. SHANKLIN METAL PRODUCTS	6	Shanklin Metal Products Co.
61. SIMMONS	4	John Simmons Company
62. S. E. SIMMONS MFG. CO.	9	Samuel E. Simmons
63. SNELL LAMP	10	Unknown; a bottom exists
64. SPRINGFIELD LAMP	6	Meyer Stein Manufacturing Co.
65. SQUARELITE	7	Shanklin Manufacturing Co.
66. STANDARD	10	John S. Cummings Co.
67. STEIN DROPPER	9	Probably Meyer Stein Mfg. Co.
68. SUN RAY	5	Dewar Manufacturing Co.
69. SURE-LITE	7	Sure-Light Acetylene Co.
70. THE BUDDY	9	Justrite Manufacturing Co.
71. UNCLE SAM	7	Justrite Manufacturing Co.
72. UNION CARBIDE	9	Unknown, strongly resembles Maple City Lamp
73. VICTOR	8	Unknown; this is the non-Justrite version
74. VICTOR	5	Justrite Manufacturing Company
75. WHAT CHEER	9	What Cheer Drill and Miners' Tool Company
76. WOLF	4	Wolf Safety Lamp Company of America
77. X-RAY (Defender)	6	Justrite Manufacturing Co.
78. X-RAY (Fulton-Imperial)	6	Unknown
79. ZAR	5	First—John Simmons Co. Second—Zar Manufacturing Co.

Appendix C

BASIC LAMPOLOGY

With all due respect, it must be stated that carbide lamps are, on the whole, strange devices. It is a source of amazement how anyone could come up with a lighting apparatus that relied upon water and smelly rocks in order to function. Yet, with nearly eighty years of reliable, if peculiar service, the dependability of these lamps is well established.

The working design behind a carbide lamp is fairly straightforward. Water, located in a chamber above the carbide, is valved through a water-feed device dependent upon gravity, and dripped on calcium carbide housed in a container underneath the water. The resulting chemical reaction between the water and carbide yields acetylene gas.

$$\underset{\text{(Calcium carbide)}}{CaC_2} + \underset{\text{(Water)}}{2H_2O} \rightarrow \underset{\text{(Calcium hydroxide)}}{Ca(OH)_2} + \underset{\text{(Acetylene)}}{C_2H_2} + \text{Heat}$$

As more acetylene is generated, pressure forces this gas through a felt filter, into a small tube, and vents it from the lamp at the burner tip or jet. Here, it is ignited, burning with a brilliant flame that is reflected by a brightly coated, concave reflector.

$$\underset{\text{(Acetylene)}}{H_2C_2} + \underset{\text{(Oxygen)}}{2O_2} + \text{spark} \rightarrow \underset{\text{(Water)}}{H_2O} + \underset{\text{(Carbon dioxide)}}{2CO_2} + \text{heat and light}$$

In theory it is very simple; in practice there can be some difficulties. However, with proper care and use, modern acetylene lamps will last a long time and give dependable service.

One of the cardinal rules of carbide lamp care is to empty the "spent carbide" from the lamp bottom when the charge is exhausted. This is easily accomplished when another fresh charge is needed, but often forgotten when the lamp is set down for the day. A couple of days wait may not be too harmful, but after several weeks, the caustic calcium hydroxide (spent carbide) can corrode the brass and "freeze-up" the lamp. Getting stuck lamps apart is no fun and sometimes the lamp may be damaged or even destroyed in the resulting frustration! One of the tried ways of doing this is to set the lamp bottom in a large (8" diameter is enough) pool of plaster of Paris, submerging the bottom to the level of the gasket. After drying thoroughly, the block of plaster often provides enough extra leverage to twist the lamp bottom apart from the water chamber. The best solution however, is to prevent this scenario and keep the lamp clean when not in use. A little oil on cleaned threads will also make for easier work.

When removing spent carbide, do not bang or knock the lamp bottom on rocks, concrete, etc. Besides possibly damaging a beautiful piece of cave flowstone, the lamp could become seriously dented or even punctured. Scrape the waste out of the bottom with a stick or piece of metal (screwdriver is good) and wash clean with water. Some prefer a little vinegar as the acid cuts the basic sludge a little better. The blackened oxide interior of the lamp bottom is simply a reaction of the brass to the carbide. Leave it; it probably provides a little protection to the underlying brass of the lamp.

There are other smart, preventative tips that can prevent lamp trouble. Periodic inspection of the felt and gasket can save unnecessary hassles later. Is the felt hard or excessively worn? Is the gasket brittle or cracked? These conditions can all lead to a problem underground and are easily and cheaply remedied.

A clean burner tip is important. Use only a tip cleaner or reamer for this job. Nails, pins, and the like do not suffice! If the tip is cracked or improperly seated, replace and reseat it. Many individuals carry a small pair of pliers just for this purpose. (The handle can also be used as a spent carbide scraper!)

A clean, bright reflector throws a brighter light than a scratched, muddy, or blackened one. A toothbrush and chrome or brass polish can do wonders; some users have even found that toothpaste is an excellent cleaner.

Keep the lighter assembly clear of mud and dirt. A small drop of oil will prevent the spring from rusting. Avoid screwing the spring too tightly against the flint. Lamp reliability is a good flint backed up by waterproofed matches.

Bumpergrips are good protection against dented lamp bottoms and do protect occasionally against burns from hot lamps. If a hot lamp persists though, check the water feed mechanism; lamps hot enough to blister are working too hard!

Regarding carbide, most commercial brands available today are of adequate quality. Oiled, miners' grade carbide is preferred (about the size of peas). Avoid buying white, dusty stuff; it has seen better days!

The lamp's water supply should be clean, the cleaner the better. A common yet unrecognized cause of "lampitis" is dirty water. Particles clog the water feed mechanism and sporadic acetylene production is the end result. One experimenter has determined the following concerning "emergency sources of liquid" for carbide lamps.[1]

URINE—will clog up lamp in 4 to 5 hours
SCOTCH—good for 2 hours
BEER—45 minutes
COLA—one hour
SALIVA—4 hours

To prevent trouble underground, most users keep some kind of spare parts kit readily available. Packed away in a bag or pocket, it can be a real "trip-saver" once underground. The typical kit might include . . .

a tip cleaner
several felts
several tips
several gaskets
a spare lamp bottom
a sturdy carbide dump bag
flint
wing nut
reamer
felt holder
felt plate
striker assembly

These last five items might be considered optional, but certainly the first seven should be mandatory. Enough cannot be said about the need for a sturdy dump bag. It should always be used; dumped calcium hydroxide is definitely detrimental to the beauty found underground and TOXIC TO LIFE FORMS BOTH IN THE CAVE AND OUTSIDE. DISPOSE OF PROPERLY.

Once the lamp is ready for use, that strange disease, lampitis, may still rear its ugly head. In the interest of ending lampitis and doctoring the lamp back to health, the following "sick lamp guide" is offered.

CARBIDE LAMP TROUBLESHOOTING GUIDE
OR
"HOW TO CURE LAMPITIS"

CURSE	CAUSE	CURE
No spark from striker assembly	Flint missing	Replace
	Spring missing	Replace
	Assembly clogged with mud, dirt, wax, etc.	Clean it
	Spring pressure too tight/loose	Loosen/tighten spring nut; replace if missing
	Assembly is damp or wet	Dry or replace assembly
	Assembly is reversed, throwing spark away from burner tip	Loosen assembly nut and reverse striker so spark is directed at burner tip
	Bad striking technique	Practice, practice (If all else fails, carry waterproofed matches)
No flame	Carbide is exhausted or missing	Clean out spent carbide and refill
	Water is exhausted or missing	Refill water chamber
	Felt is wet, brittle, dirty, or extremely worn	Clean gasket seating area and replace (Mud has been used in emergencies as gasket material, but will dry rapidly, crack, and fall away)
	Cracks or holes in lamp	Repair lamp with solder or replace; change bottoms (Radiator sealer works poorly here! Mud is not much better!)
	Lamp bottom is not seated properly	Clean and oil lamp threads. Carefully rethread.
	Tip is clogged or blocked	Clean tip with tip cleaner. To widen orifice, use tip reamer
	Gas tube to burner tip is plugged or blocked	Remove tip, felt holder, felt, felt plate. Clean and brush out tube
	Water door vent is clogged (This may prevent water from dropping through valve to carbide)	Clean
	Poor striking technique (assuming there is a spark)	Practice, practice. Cup hand over reflector for 15 seconds. Just before striking, open hand slightly, permitting O_2 to mix with C_2H_2. Strike in a deliberate motion
	Water valve is clogged, broken	Increase water flow slowly. If malfunction persists, clean spent carbide from valve end and recheck water flow. Blow through water tank to clear stubborn debris. Repair/replace water valve/lamp if broken

CURSE	CAUSE	CURE
Water bubbles exiting from tip	Soaked felt and gas tube	Replace/dry felt. Filter on cigarette can sponge water out of gas tube to burner tip. (In a pinch, these can act as temporary felts.)
	Free-diving sumps	Either go electric or blast a way around
Flame coming from around gasket	Lamp is threaded incorrectly	Carefully rethread lamp
	Gasket is old, cracked, brittle	Replace gasket. Clean gasket seating area
Flame coming from water door vent	Gas is being blocked from burner tip and being shunted back through water feed into water tank and subsequently ignited	Ream burner tip, replace felt, clean foreign debris from gas tube to burner tip.
Sputtering flame	Tip is cracked, misfitted	Replace tip
	Water in burner tip tube, felt is wet	Dry out tube, replace felt
Tilted or angled flame	Cracked burner tip	Replace burner tip
	Dirty burner tip	Clean or ream
Weak, irregular flame	Water supply depleted	Refill water supply
	Carbide supply depleted	Refill lamp bottom with carbide
	Burner tip partially plugged	Clean, ream, or replace burner tip
	Wet felt	Dry or replace
	Cracks in lamp; acetylene escapes and may ignite	Repair cracks or replace lamp
	Gasket is old, cracked, or missing	Clean gasket seating area and replace with new gasket
	Using dirty water	Flush water feed valve, replace with clean water
Excessive flame (temporary)	Too much water being valved onto carbide	Turn off water valve and wait for excess gas to be vented
	Burner tip has been reamed excessively, oriface widened (note very yellow flame)	Replace tip
Excessive flame (spasmodic)	Sheer force of acetylene venting from burner tip blows out existing flame	Cut back water supply; wait for excess gas to vent
	The above situation with a plugged water door vent	Unplug vent, cut back water supply. Wait for excess gas to vent.
Excessive flame (continuous and uncontrollable)	Broken water feed device; wide open water feed	Repair valve, valve seat or replace lamp, whichever is cheaper
Bubbling in water tank (often noted in normally functioning lamps)	Excess acetylene being shunted through water valve and percolating through water tank	Cut back acetylene generation by slowing water feed
Very hot lamp, particularly bottoms	Too much acetylene being generated	Slow water feed to carbide
	Missing bumper grip	Replace
Good flame but poor illumination	Dirty reflector	Clean with toothbrush and brass cleaner, etc.
	Eyes	Remove sunglasses
	Cave character	Use a brighter light if cave walls, ceiling, etc. are colored unusually dark

Needless to say, it is impossible to account here for every possible situation that might lead to lamp failure. For example, if there still is no flame after checking all other possibilities, it may be that there is little oxygen in the area where the lamp is trying to burn. . . . a good time to exit the cave. Or possibly, instead of putting carbide into the bottom, suppose peanuts or raisins got in there by accident!

With a good working knowledge of the lamp and its various parts, a well supplied parts kit, and experience in repairing/recharging the lamp in darkness, the user should encounter few if any, unsolvable cases of "lampitis."

Bibliography

Articles

Anon., "Accident Report," *NSS News*, Vol. 24, No. 5.

———, "Acetylene Gas for Mine Lighting," *The Engineering and Mining Journal*. October 4, 1902.

———, "Acetylene Lamps for Mine Use," *Acetylene Journal*. October 1906.

———, "Acetylene Lamps for Miners' Use," *The Colliery Guardian*. February 23, 1906.

———, "Acetylene-Lamp Improvements," *The Engineering and Mining Journal*. October 28, 1916.

———, "Acetylene Lighting Beneath the Earth's Surface," *Acetylene Journal*. January 1911.

———, "Acetylene Underground," *The Engineering and Mining Journal*. August 14, 1897.

———, "An Acetylene Miner's Lamp," *Mining and Engineering World*. July 1, 1911.

———, "A New Mine Lamp for Using Acetylene Gas," *Engineering and Mining Journal*. October 12, 1901.

———, "Anleitung fur Acetylenverbraucher," *Acetylen in Wissenschaft und Industrie*. March 1916.

———, "Arnold Carbide Candle," *The Engineering and Mining Journal*. January 3, 1914.

———, "Awards Announced for Acetylene Mine Lamps," *Acetylene Journal*. August 1915.

Bailey, C. E. "Acetylene Mining Lamps," *Acetylene Journal*. November 1904.

"Baldwin Acetylene Bicycle Lamp," *Scientific American*. June 23, 1900.

Baldwin, F. E. "Miners' Acetylene Lamp," *Mining and Engineering World*. July 11, 1914.

Brown, Paul. "Conversion of a Carbide Lamp to Carry the Generator on the Belt," *The Netherworld News*. Vol. 8, No. 8, 1960.

Bullard, E. W. "Introduction of the Acetylene Mine Lamp on the Pacific Coast," *Acetylene Journal*. October 1916.

———: "History of the Carbide Mine Lamp from the Earliest Days to the Present Time," *Acetylene Journal*. December 1916.

———: "Miner's Carbide Lamps," *Acetylene Journal*. January 1917.

Burrows, Robert P. "Illumination of Mines," *Transactions of the American Institute of Mining Engineers*. Vol. 54, 1916.

"Candles and Carbide Lamps," *Journal of Acetylene Lighting*. November 1917.

"Cane Torches as Cave Illumination," *NSS News*. Vol. 24, No. 3.

"Carbide Versus Electric Lamps," *The Engineering and Mining Jounral*. November 13, 1926.

Chance, E. M. "The Ability of the Acetylene Flame," *The Colliery Engineer*. November 1913.

————. "Portable Miners' Lamps," *Transactions of the American Institute of Mining Engineers*. Bulletin No. 22, February 1917.

Clemmer, G. S. "Short History of American Acetylene Mine Lamps," *NSS News*. June 1982.

Cremer, R. "Acetylene Lamps in Mines," *The Mining Journal*. March 22, March 29, and April 5, 1913.

Deszeny, Julius. "Acetylene as used in a mine where 6,000 men are employed," *Acetylene Journal*. January 1912.

————: "Acetylene in Underground Mines," *Montan Zeitung*. August 1912.

"Die-Cast Aluminum Lamp Made in New Large Size," *The Engineering and Mining Journal Press*. October 10, 1925.

"Electric Lighting in Mines," *The Engineering and Mining Journal*. Vol. 48, 1889.

"Every-Ready Acetylene Lamp," *The Engineering and Mining Journal*. January 3, 1914.

Frank, G. "Experiments with Acetylene Lamps for Pit Work," *Colliery Guardian*. March 21, 1902.

"German use of Acetylene Lamps," *The Engineering and Mining Journal*. November 25, 1899.

Harrison, Gene. "Everything You Should Know About Plastic Carbide Lamps (But were afraid you'd find out!)," *The Potomac Caver*. April 1972.

Hodgson, L. H. "Acetylene Safety Lamps," *Transactions of the Institute of Mining Engineers*. Vol. 32, 1906-07.

Holland, A. "Acetylene as a Mine Illuminant," *Iron and Coal Trades Review*. April 28, 1899.

Huston, George. "Acetylene Mine Lamps," *Mining Science*. September 26, 1912.

Iles, A. "A Carbide Incident in Agen Allwedd," *Speleologist*. February/March 1966.

Jackson, George. "Russell Trall Neville Expedition in Old Salts Cave, KY," *Journal of Spelean History*. Vol. 2, No. 3, 1969.

Kellogg, L. O. "The Acetylene Lamp Underground," *The Engineering and Mining Journal*. December 21, 1912.

Kissell, Fred. "Modified Carbide Lamp," *The Netherworld News*. Vol. 8, No. 11.

Lee, Dana W. "The National Geographic Society Expedition of Carlsbad Caverns, New Mexico," *Journal of Spelean History*. Vol. 1, No. 3.

Leeds, F. H. "Acetylene Lamps in Mines," *Acetylene Journal*. October 1905.

Lenhart, Warren G. "Acetylene versus Candles," *Mining and Scientific Press*. September 30, 1916.

"Letang and Serpollet's Acetylene Lamp," *Scientific American Supplement*. May 1, 1897.

Luxton, W. H. "Letter to Editor," *The Colliery Engineer*. September 1913.

Mapother, W. L. "Acetylene in Colossal Cavern," *Acetylene Journal*. January 1905.

Matthews, Larry. "Photography by Carbide Light," *Nittany Grotto Newsletter*. Vol. 5, No. 5, (See also, SpeleoDigest-1957.)

Maurice, William. "Acetylene Mine Lamps," *Transactions–Institution of Mining Engineers*. Vol. 53, 1916-17.

————. "Chance Acetylene Safety Lamp," *Coal Age*. Vol. 14, 1918.

————. "The Wolf-Pokorny and Wiede Acetylene Safety Lamps," *Transactions–Institution of Mining Engineers*. Vol. 59, 1920.

Meador, Tom. "Why Not Dump It Here?" *Texas Caver*. Vol. 10, No. 11.

Menner, H. M. "Acetylene Lamps in Coal Mines," *The Colliery Engineer*. August 1913.

Mitick, Ellsworth C. "A History of Mining Machinery Manufacture," M.S. thesis published in *Colorado Magazine*. Vol. 25, No. 2, March 1948.

Morehead, J. T. and de Chalmont, G. "Manufacture of Calcium Carbide," *Journal of the American Chemical Society*. April 1896.

Morley, Frederick H. "Acetylene Lamps for Metal Mines," *Mining and Scientific Press*. April 11, 1914.

Morrison, A. Cressy. "Better Light and Air for Miners," *The Mining World*. December 12, 1908.

————. "Acetylene Mine Lighting," *Mining Science*. April 1915.

"New Acetylene Mine Lamps," *The Engineering and Mining Journal*. June 27, 1914.

Oke, A. Livingstone. "A Home-Made Acetylene Lamp," *The Engineering and Mining Journal*. September 6, 1913.

Owings, C. W. "Some Mine-Lighting Practices in the United States," *Transactions of the Illuminating Engineering Society*. Vol. 29, No. 1, January 1934.

Parsons, F. W. "Acetylene Lamps for Mines," *The Engineering and Mining Journal*. July 21, 1906.

Powers, John. ed. "Subterranean Sun," Vol. 6, No. 4, April 1976.

Rice, Tom. "Weight for Hardhat Balances Carbide Lamp," *The California Caver*. Vol. 14, No. 1, 1962.

"Safety Lamps," *Coal Age*. Vol. 1, 1911.

Schroeder, R. B. "Lamp Mounting System," *Nittany Grotto Newsletter*. Vol. 11, No. 4, 1963.

"Self-Cleaning Attachment for Burners of Carbide Lamps," *The Engineering and Mining Journal*. September 3, 1921.

Shoemaker, G. M. "Letter to Editor," *The Colliery Engineer*. September 1913.

Smith, E. E. "Acetylene as an Illuminant in Mines," *Mining and Engineering World*. December 20, 1913.

Stephenson, William J. "The Founding of the N.S.S.," *Journal of Spelean History*. Vol. 2, No. 4.

Stade, John. "Calcium Carbide," *The Underground*.

Vol. 8, No. 2, 1965.

"Surveying Reflector for Acetylene Lamp," *The Engineering and Mining Journal*. February 21, 1914.

"The Baldwin Acetylene Lamp," *The Engineering and Mining Journal*. September 15, 1900.

"The Electric Light at the Nunnery Colliery," *Colliery Guardian*. 1880.

"The Milburn Establishment, Baltimore," *Acetylene Journal*. Feburary 1914.

"The Wolf Acetylene Safety Lamp," *Mines and Minerals*. November 1906.

"Turning Darkness Into Day," *Acetylene Journal*. September 1903.

"Untitled," *Acetylene Gas Journal*. February 1900, p. 94.

"Untitled," *The Engineering and Mining Journal*. July 26, 1902.

"Untitled," *Michigan Miner*. November 1902, p. 27.

"Untitled," *Acetylene Journal*. September 1903, p. 90.

"Untitled," *Mine and Minerals Directory*. February 1909, p. 5.

"Untitled," *Mining Science*. October 27, 1910.

Venable, F. P. "Account of the First Production of Calcium Carbide and Acetylene in U.S.," *American Manufacturing and Iron World*. December 16, 1898.

Willis, Charles F. "Advantages of Acetylene Lighting for Mines," *Journal of Acetylene Lighting*. September 1917.

Willson, T. L. and Suckert, J. J. "Carbides and Acetylene Commercially Considered," *Journal of Franklyn Institute*. May 1895.

"Wolf Lamp used in Newhouse Tunnel," *Mining and Scientific Press*. May 18, 1907.

"World-Famed Natural Cave in Tennessee Beautifully Illuminated with Acetylene," *Acetylene Journal*. August 1914.

Young, Chuck. "Improved Tip Cleaner," *The Potomac Caver*. June 1971.

Newspapers

Oshkosh Daily Northwestern—Wisconsin.Illinois State Journal, 1926-1930.

Books

Bowles, J. D. *Acetylene for Lighting Country Homes*, University of Missouri, March 1910.

Bingham, Charles. *The Manufacture of Carbide of Calcium*. London, 1916.

Halliday, Dr. W. R. *American Caves and Caving*. New York, 1974.

Hovey, Rev. Horace. *The Mammoth Cave of Kentucky*. Louisville, 1912.

Lewes, Vivian B. *Acetylene,* New York, 1900.

Nouveau Guide Practique de L'usager D'acetylene. Paris. 1914.

Paul, James W. *Notes on Miners' Carbide Lamps*. Miners Circular No. 18, U.S. Bureau of Mines, Washington, D.C., 1915.

Pohs, Henry A. *Early Underground Mine Lamps*. Tucson, Arizona, 1974.

Pond, George Gilbert. *Calcium Carbide and Acetylene*. State College, Pennsylvania, 1908.

Zern, E. N. ed. *Coal Miners' Pocketbook*. 12th Edition, New York, 1928.

Business—City Directories

Alton, Illinois
Ashland, Kentucky
Baltimore, Maryland
Cincinnati, Ohio
Chicago, Illinois
Denver, Colorado
Evansville, Indiana
Huntington, West Virginia
Jersey City, New Jersey
Monmouth, Illinois
Monongahela, Pennsylvania
New York, New York
Oshkosh, Wisconsin
Pittsburgh, Pennsylvania
San Francisco, California
Scranton, Pennsylvania
Springfield, Illinois
Waterbury, Connecticut
Wilkes-Barre, Pennsylvania

Collectors' Newsletters, Pamphlets, Catalogs, etc.

Belknap Hardware and Manufacturing Company–Catalog. Louisville. 1945.

Brilliant Search-lite Catalog. 1918.

Bureau of Mines–Bulletins Nos. 57, 227, 359. Department of Interior. Washington, D.C.

Coal Field Directory and Mining Catalogues. Keystone Consolidated Publishing. Pittsburgh. 1915-27.

Donald Davis Papers. Parachute, Colorado.

Justrite Manufacturing Company Catalogs. Nos. 1, 5, 7, 10.

Kouts, Paul. *Miner's Carbide Lamp Reference*. Vol. 1-11 continued., December 1981-. Albuquerque, New Mexico.

Montgomery Ward Catalog–1926.

Pohs, Henry, ed. *Underground Lamp Post*. 1968-. Denver.

Sears and Roebuck Catalog–1922.

Thomas Register of Manufacturers. Thomas Publishing Company, New York. 1905-.

Whidby, Jim, ed. *Underground Lighting Section of the N.S.S.* Vol. 1.-, 1983-. Knoxville, Tennessee.

Wilson, Wendell E. *The Nearly Complete Bottom Guide for Carbide Cap Lamps*. Second Edition. Tucson, Arizona. 1982.

Index